HEAT AND MASS TRANSFER DATA BOOK

(Eighth Edition)

C P Kothandaraman
Former Head
Department of Mechanical Engineering
PSG College of Technology, Coimbatore
Tamil Nadu

S Subramanyan
Former Principal
PSG College of Technology
and
Vice Chancellor, Bharathiar University
Coimbatore, Tamil Nadu

NEW AGE TECHNO PRESS

An Imprint of

NEW AGE INTERNATIONAL (P) LIMITED, PUBLISHERS
New Delhi • Bangalore • Chennai • Cochin • Guwahati
Hyderabad • Kolkata • Lucknow • Mumbai
Visit us at www.newagepublishers.com

Copyright © 2014, 2010, 2007, 2004, 1989, New Age International (P) Ltd., Publishers
Published by New Age International (P) Ltd., Publishers
First Edition: 1989
Eight Edition: 2014
Reprint: 2015

All rights reserved.
No part of this book may be reproduced in any form, by photostat, microfilm, xerography, or any other means, or incorporated into any information retrieval system, electronic or mechanical, without the written permission of the publisher.

BRANCHES

- **Bangalore** 37/10, 8th Cross (Near Hanuman Temple), Azad Nagar, Chamarajpet, Bangalore-560 018
  Tel.: (080) 26756823, **Telefax:** 26756820, **E-mail:** bangalore@newagepublishers.com

- **Chennai** 26, Damodaran Street, T. Nagar, Chennai-600 017
  Tel.: (044) 24353401, **Telefax:** 24351463, **E-mail:** chennai@newagepublishers.com

- **Cochin** CC-39/1016, Carrier Station Road, Ernakulam South, Cochin-682 016
  Tel.: (0484) 2377303, **Telefax:** 4051303, **E-mail:** cochin@newagepublishers.com

- **Guwahati** Hemsen Complex, Mohd. Shah Road, Paltan Bazar, Near Starline Hotel, Guwahati-781 008
  Tel.: (0361) 2513881, **Telefax:** 2543669, **E-mail:** guwahati@newagepublishers.com

- **Hyderabad** 105, 1st Floor, Madhiray Kaveri Tower, 3-2-19, Azam Jahi Road, Near Kumar Theater Nimboliadda, Kachiguda, Hyderabad-500 027, **Tel.:** (040) 24652456, **Telefax:** 24652457
  E-mail: hyderabad@newagepublishers.com

- **Kolkata** RDB Chambers (Formerly Lotus Cinema) 106A, 1st Floor, S.N. Banerjee Road, Kolkata-700 014
  Tel.: (033) 22273773, **Telefax:** 22275247, **E-mail:** kolkata@newagepublishers.com

- **Lucknow** 16-A, Jopling Road, Lucknow-226 001
  Tel.: (0522) 2209578, 4045297, **Telefax:** 2204098, **E-mail:** lucknow@newagepublishers.com

- **Mumbai** 142C, Victor House, Ground Floor, N.M. Joshi Marg, Lower Parel, Mumbai-400 013
  Tel.: (022) 24927869, **Telefax:** 24915415, **E-mail:** mumbai@newagepublishers.com

- **New Delhi** 22, Golden House, Daryaganj, New Delhi-110 002
  Tel.: (011) 23262368, 23262370, **Telefax:** 43551305, **E-mail:** sales@newagepublishers.com

ISBN: 978-81-224-3595-5

₹199.00

C-14-12-8103

Printed in India at Repro India Ltd., Mumbai.
Typeset at In-house, Delhi.

PUBLISHING GLOBALLY
NEW AGE INTERNATIONAL (P) LIMITED, PUBLISHERS
7/30 A, Daryaganj, New Delhi-110002
Visit us at www.newagepublishers.com

To the memory of

PROF. G.R. DAMODARAN

Founder Principal, PSG College of Technology
and
Vice-Chancellor, University of Madras (1978–1981)

PREFACE TO THE EIGHTH EDITION

The aim of this book is to present to the students, teachers and practising engineers as a comprehensive collection of various material property data and formulae in the fields of heat and mass transfer. The material is organized in such a way that a reader who has gone through engineering curriculum could easily use the formulae and data presented in heat transfer calculations. Hence, this compilation, intended as an adjunct to a standard text, serves as a quick reference for both the students and practising engineers in the filed.

Considerable space has been devoted to property values of materials – solids, liquids and gases – that are commonly used in heat transfer situations. Property values of various materials at different temperatures are also given. They have been arranged in alphabetical order for quick reference with additions to the earlier edition.

The tabular form of presentation of the formulae with symbols, units, the limitations and restrictions in the use of empirical relationships explained alognside, has been retained, as several favourable comments have been received on this point. The empirical formulae and charts have been selected with a view to presenting information with clarity to facilitate straightforward use.

As far as possible, suggestions received since the appearance of the seventh edition have been incorporated in the new edition. A number of charts have been redrawn or added to improve clarity. To avoid ambiguity, property values have been tabulated fully without multiplication factors in the column heading.

The presentation of convection has been enlarged, taking into account the recent publications. Properties for solar application, information for the design of external finned heat exchangers and extensive emissivity values are noteworthy addition. Besides in this edition, transport property data for seventeen heat-pipe materials, steam and some popularly used refrigerants are included. The presentation of data and equations are presented only in SI units. However, appropriate conversion factors have been provided extensively for those who want to use MKS system of units.

The book is an outcome of the authors' teaching the subject to the undergraduate students at the PSG College of Technology for over 35 years. We hope that it serves in training the students with problem of purposeful and practical design situations. The data book has been favourably received by several Indian universities and many of them have permitted its use in the university examinations for reference.

We gratefully acknowledge the encouragement given in our efforts by Professor G.R. Damodaran to who we dedicate this book. Our thanks are due to our former Principals, Dr. R. Subbayyan, Dr. K. Venkatraman, and Dr. A. Shanmugasundaram for the facilities afforded to us.

Coimbatore

C.P. Kothandaraman
S. Subramanyan

CONTENTS

Preface .... *(vii)*

PROPERTY VALUES

    Metals 1
    Alloys 9
    Non Metallic Solids 11
    Common Materials 13
    Insulating, Building and Other Materials 15
    Liquids 22
    Liquid Metals 32
    Gases & Vapours 34

CONDUCTION

    Steady State 44
    Steady State with Heat Generation 48
    Fins 50
    TRANSIENT :
        Lumped Parameter System 58
        Semi Infinite Solid 59
        Infinite Solids 64
        Periodic Heat Flow 76
        Freezing and Melting 77

RADIATION

    Radiation 82
    Heat Transfer between Surfaces 84
    Gas Radiation 105

CONVECTION

    FORCED :
        External Flow 113

  Internal Flow 124
  Packed Beds 133
 Free Convection 135
 Boiling 148
 Condensation 149

HEAT EXCHANGER

 Regenerative Type 152
 Storage Type 169
 Transverse Fin 171

MASS TRANSFER

 Basic Equation for Diffusion Coefficient 174
 Diffusion Mass Transfer 175
 Convective Mass Transfer 176
 Humidification 178
 Diffusion Coefficients, Schmidt number and Henry's Constant 180

APPENDIX

 Table 1: HP1 Properties of Heat Pipe Fluids – Lithium, Potassium,
     Sodium and Mercury 187

 Table 2: HP2 Properties of Heat Pipe Fluids – Caesium, Flutec PP9,
     High Temperature Organic (Diphenyl – Diphenyl Oxideutectic)
     and Flutec PP2 188

 Table 3: HP3 Properties of Heat Pipe Fluids – Heptane, Acetone,
     Methanol and Ammonia 189

 Table 4: HP4 Properties of Heat Pipe Fluids – Nitrogen, Helium,
     Ethanol, Water and Pentane 190

 Table 5: Variation of Thermal Conductivity and Viscosity of
     R12, Dichlorodifluoromethane and R22, Chlorodifluoromethane ... 191

 Table 6: Variation of Thermal Conductivity and Viscosity of
     R134a, Tetrafluoroethane and R717, Ammonia 192

 Table 7: Variation of Thermal Conductivity and Viscosity of
     R744, Carbon Dioxide and R32, Difluoromethane 193

 Table 8: Transport Properties of Water and Steam
     at Saturation Conditions 194

 Table 9: Thermal Conductivity, k, of Steam and Water
     at Various Pressures and Temperatures 195

 Table 10: Dynamic Viscosity of Steam and Water at Various
     Pressures and Temperatures 196

 Table 11: Prandtl Number of Steam and Water
     at Various Pressures and Temperatures 197

UNIT CONVERSION CONSTANTS

REFERENCES

PSYCHROMETRIC CHARTS

HEAT AND MASS TRANSFER DATA BOOK

PROPERTY VALUES OF METALS AT 20°C OR AS INDICATED

Metal	Density ρ kg/m^3	Thermal Diffusivity $\alpha = k/\rho c$ m^2/s	Specific Heat c J/kgK	Thermal Conductivity k W/mK
Aluminium, Pure	2707	84.18×10^{-6}	896	204.2
Aluminium alloy 2024-T6 (4.5% Cu, 1.5% Mg, 0.6% Mn) at 27°C	2770	73.00×10^{-6}	875	177.0
Aluminium alloy 195, cast, (4.5% Cu) at 27°C	2790	68.20×10^{-6}	883	168.0
Aluminium Bronze, (95% Cu, 5% Al)	8666	23.30×10^{-6}	410	83.0
Al-Cu, Duralumin (94.96% Al, 3-5% Cu, trace Mg.)	2787	66.76×10^{-6}	883	164.5
Al-Mg-Si (97% Al, 1% Mg, 1% Si, 1% Mn.)	2707	73.11×10^{-6}	892	117.0
Al-Si (Alusil) (78.8% Al, 20.22% Si)	2627	71.72×10^{-6}	854	161.0
Al-Si (Silumin) (86.5% Al, 12.5% Si, 1% Cu)	2659	59.33×10^{-6}	867	136.8
ARMCO IRON (99.75% pure) at 27°C	7870	20.70×10^{-6}	447	72.7
Brass, (70% Cu, 30% Zn)	8522	34.12×10^{-6}	385	110.7
Brass, Red (85% Cu, 9% Sn, 6% Zn)	8714	18.04×10^{-6}	385	60.6
Bronze (75% Cu, 25% Sn)	8666	8.59×10^{-6}	343	25.9
Bronze, Commercial, (90% Cu, 10% Al) at 27°C	8800	14.00×10^{-6}	420	52.0
Bronze, Phosphor (87% Cu, 11% Sn) at 27°C	8780	17.00×10^{-6}	355	54.0
Carbon steel, Plain carbon, (Mn ≤ 1%, Si ≤ 0.1%) at 27°C	7854	17.70×10^{-6}	434	60.5
Carbon steel, AISI 1010	7832	18.80×10^{-6}	434	63.9
Carbon-Silicon Steel, (Mn ≤ 1%, 0.1% < Si ≤ 0.6%) at 27°C	7817	14.90×10^{-6}	446	51.9
Carbon-Manganese-Silicon Steel (1% < Mn ≤ 1.65%, 0.1% < Si ≤ 0.6%) at 27°C	8131	11.60×10^{-6}	434	41.0
Carbon Steel (Approximately 0.5%C)	7833	14.74×10^{-6}	465	53.6
Carbon Steel (Approximately 1.0%C)	7801	11.72×10^{-6}	473	43.3

$1 \, W/mK = 0.86 \, kcal/m \, hr \, °C$

$1 \, J/kg \, K = 238.9 \times 10^{-6} \, kcal/kg \, °C$

PROPERTY VALUES OF METALS AT 20°C OR AS INDICATED (Contd.)

Metal	Density ρ kg/m³	Thermal Diffusivity $\alpha = k/\rho c$ m²/s	Specific Heat c J/kgK	Thermal Conductivity k W/mK
Carbon Steel, (Approximately 1.5% C)	7753	9.70×10^{-6}	486	36.3
Cast iron (< 4% C)	7272	17.03×10^{-6}	420	52.0
Chromium Steel (low) $\frac{1}{2}$ Cr – $\frac{1}{4}$ Mo – Si (0.18% C, 0.65% Cr, 0.23% Mo, 0.6% Si) at 27°C	7822	10.90×10^{-6}	444	37.7
Chromium Steel (low) 1Cr – $\frac{1}{2}$ Mo, (0.16% C, 1% Cr, 0.54% Mo, 0.39% Si) at 27°C	7858	12.20×10^{-6}	442	42.3
Chromium Steel (low) 1Cr – V, (0.2% C, 1.02% Cr, 0.15% V) at 27°C	7836	14.10×10^{-6}	443	48.9
Chrome Steel, (1% Cr)	7865	16.65×10^{-6}	461	60.6
Chrome Steel, (5% Cr)	7833	11.10×10^{-6}	461	39.8
Chrome Steel (20% Cr)	7689	6.35×10^{-6}	461	22.5
Chrome-Nickel Steel (15% Cr, 10% Ni)	7865	5.26×10^{-6}	461	19.1
Chrome-Nickel Steel (18% Cr, 8% Ni), (V2A)	7817	4.44×10^{-6}	461	16.3
Chrome-Nickel Steel (20% Cr, 15% Ni)	7833	4.15×10^{-6}	461	15.1
Chrome-Nickel Steel (25% Cr, 20% Ni)	7866	3.61×10^{-6}	461	12.8
Constantan (60% Cu, 40% Ni)	8922	6.12×10^{-6}	410	22.7
Copper, Pure	8954	112.34×10^{-6}	383	386.0
German Silver (62% Cu, 15% Ni, 22% Zn)	8618	7.33×10^{-6}	394	24.9
Invar (36% Ni)	8137	2.86×10^{-6}	461	10.7
Inconel x 750 (71% Ni, 15% Cr, 6.7% Fe) at 27°C	8510	11.70×10^{-6}	439	3.1
Iron, Pure	7897	20.34×10^{-6}	452	72.7

$1 \, W/mK = 0.86 \, kcal/m \, hr \, °C$

$1 \, J/kg \, K = 238.9 \times 10^{-6} \, kcal/kg \, °C$

PROPERTY VALUES OF METALS AT 20°C OR AS INDICATED (Contd.)

Metal	Density ρ kg/m^3	Thermal Diffusivity $\alpha = k/\rho c$ m^2/s	Specific Heat c $J/kg\,K$	Thermal Conductivity k W/mK
Lead	11393	23.43×10^{-6}	130	34.7
Magnesium, Pure	1746	97.08×10^{-6}	1013	171.3
Mg-Al (Electrolytic) (6–8% Al, 1.2% Zn)	1810	36.05×10^{-6}	1005	65.7
Manganese Steel (5% Mn)	7849	4.83×10^{-6}	460	22.0
Molybdenum	10220	47.90×10^{-6}	251	122.8
Nickel ; Pure (99.9%)	8906	22.66×10^{-6}	446	90.0
Nickel-Chrome (90% Ni, 10% Cr)	8666	4.44×10^{-6}	444	17.2
Nicrome (80% Ni, 20% Cr)	8314	3.43×10^{-6}	444	12.6
Nickel Steel (20% Ni)	7993	5.26×10^{-6}	461	19.1
Nickel Steel (40% Ni)	8169	2.79×10^{-6}	461	10.4
Nickel Steel (80% Ni)	8618	8.72×10^{-6}	461	34.7
Platinum, Pure at 27°C	21450	25.10×10^{-6}	133	71.6
Platinum-Rhodium alloy, (60% Pt, 40% Rh) at 27°C	16630	17.40×10^{-6}	162	47.0
Silver, Purest	10524	170.04×10^{-6}	234	419.0
Silver, Pure (99.9%)	10525	165.63×10^{-6}	234	406.8
Silicon Steel (5% Si)	7417	5.55×10^{-6}	460	19.0
Stainless Steels, AISI 302 at 27°C	8055	3.91×10^{-6}	480	15.1
Stainless Steels, AISI 304 at 27°C	7900	3.95×10^{-6}	477	14.9
Stainless Steel AISI 316 at 27°C	8238	3.48×10^{-6}	468	13.4
Stainless Steel AISI 347 at 27°C	7978	3.71×10^{-6}	480	14.2
Tin, Pure	7304	38.84×10^{-6}	226	64.1
Tungsten	19350	62.71×10^{-6}	134	162.7
Tungsten Steel, (1% W)	7913	18.58×10^{-6}	448	65.7
Tungsten Steel, (5% W)	8073	15.25×10^{-6}	435	53.6
Tungsten Steel, (10% W)	8314	13.91×10^{-6}	419	48.5
Wrought Iron (C < 0.5%)	7849	16.26×10^{-6}	461	58.9
Zinc, Pure	7144	41.06×10^{-6}	385	112.1

$1\ W/mK = 0.86\ kcal/m\ hr\ °C$

$1\ J/kg\ K = 238.9 \times 10^{-6}\ kcal/kg\ °C$

PROPERTY VALUES OF METALS AT 20°C OR AS INDICATED

Element		Density ρ kg/m^3	Thermal Diffusivity $\alpha = k/\rho c$ m^2/s	Specific Heat c J/kg K	Thermal Conductivity k W/mK
Beryllium		1850	68.30×10^{-6}	1750	205.0
Bismuth	at 27°C	9780	6.59×10^{-6}	122	7.9
Boron		2500	10.90×10^{-6}	1047	28.6
Cadmium		8650	48.50×10^{-6}	231	97.0
Chromium		7160	29.00×10^{-6}	440	69.8
Cobalt		8862	20.24×10^{-6}	389	69.8
Germanium	at 27°C	5360	34.70×10^{-6}	322	59.9
Gold		19300	126.90×10^{-6}	129	317.0
Iridium		22500	50.30×10^{-6}	130	147.0
Lithium		534	42.70×10^{-6}	339.1	68.6
Manganese		7290	2.20×10^{-6}	486	7.8
Niobium	at 27°C	8570	23.60×10^{-6}	265	53.7
Palladium	at 27°C	12020	24.50×10^{-6}	244	71.8
Platinum		21450	25.00×10^{-6}	133	69.8
Potassium		860	161.86×10^{-6}	741	100.0
Rhodium		12450	48.60×10^{-6}	248	150.0
Sodium		971	113.60×10^{-6}	1206	109.3
Silicon		2330	93.40×10^{-6}	703	153.0
Thorium		11700	39.11×10^{-6}	118	54.0
Titanium		4500	8.00×10^{-6}	611	15.12
Tungsten	at 27°C	19300	68.5×10^{-6}	132	174.0
Uranium		19070	12.70×10^{-6}	113	27.4
Vanadium		6100	10.30×10^{-6}	502	34.9
Zirconium		6570	12.80×10^{-6}	272	22.7

1 W/mK = 0.86 kcal/m hr °C   *1 J/kg K = 238.9 $\times 10^{-6}$ kcal/kg °C*

VARIATION OF THERMAL CONDUCTIVITY OF METALS WITH TEMPERATURE (W/mK)

Metal	-100	0	100	200	300	400	600	800	1000	1200
Aluminium, pure	214.6	202.5	206.0	214.6	228.2	249.2	—	—	—	—
Al-Cu Duralumin, 94-96% Al, 3-5% Cu Trace Mg.	126.5	159.2	181.8	193.9	—	—	—	—	—	—
Al-Mg-Si, 97% Al, 1% Mg, 1% Si, 1% Mn	—	175	189	204	—	—	—	—	—	—
Al-Si (Alusil) 78-80 % Al, 20-22% Si	144	157	168	175	178	—	—	—	—	—
Al-Si (Silumin, Cu bearing 86.5%, Al, 12.5% Si, 1% Cu)	119.3	136.8	143.6	152.2	161.0	—	—	—	—	—
Brass, 70% Cu, 30% Zn	88.3	—	128.1	143.9	147.1	147.1	—	—	—	—
Bronze, 66.4% Cu, 31.1% Zn, 2.3% Mn	69.78	73.62	73.63	77.92	82.57	88.38	—	—	—	—
Chrome Steel, 1% Cr	—	62.3	55.4	51.9	46.8	41.5	36.3	32.9	32.9	—
Chrome Steel, 5% Cr	—	39.8	38.0	36.3	36.3	32.9	29.4	29.4	29.4	—
Chrome Steel, 20% Cr	—	22.4	22.4	22.4	22.4	24.2	24.2	25.9	29.4	—
Chrome Nickel Steel, 18% Cr, 8% Ni (V2A)	—	16.3	17.3	17.3	19.1	19.1	22.5	27.0	31.2	—
Constantan 60% Cu, 40% Ni	20.8	—	22.1	25.9	—	—	—	—	—	—
Copper, Pure	406.7	386.0	379.0	373.9	368.7	363.6	353.1	—	—	—
German Silver, 62% Cu, 15% Ni, 22% Zn	19.2	—	31.2	39.8	45.0	48.5	—	—	—	—
Iron, Pure	86.5	72.7	67.5	62.3	55.4	48.5	39.8	36.3	34.7	36.3

Temperature, °C

$1\ W/mK = 0.86\ kcal/m\ hr\ °C$

VARIATION OF THERMAL CONDUCTIVITY OF METALS WITH TEMPERATURE (W/mK) *(Contd.)*

Metal	−100	0	100	200	300	400	600	800	1000	1200
Lead	36.9	35.1	33.4	31.4	29.8	—	—	—	—	—
Magnesium, Pure	178.3	171.3	167.8	162.8	157.5	—	—	—	—	—
Mg-Al (Electrolytic) 6-8% Al, 1,2% Zn	—	52.0	62.3	74.4	83.2	—	—	—	—	—
Molybdenum	138.4	124.6	117.5	114.3	111.0	109.1	105.6	102.1	98.6	91.8
Nickel, Pure (99.9%)	103.9	93.5	83.2	72.7	64.1	58.0	—	—	—	—
Nickel-chrome 90% Ni, 10% Cr	—	17.1	18.8	20.9	22.9	24.7	—	—	—	—
Nickel-chrome 80% Ni, 20% Cr	—	12.3	13.3	15.6	17.1	18.0	22.5	—	—	—
Red Brass, 85% Cu, 9% Sn, 6% Zn	—	58.9	70.9	—	—	—	—	—	—	—
Silver, Pure (99.9%)	418.9	410	415	373.9	362.3	360.5	—	—	—	—
Steel (Carbon Steel)										
Approx 0.5% C	—	55.4	51.9	48.5	45.0	41.5	34.7	31.2	29.4	31.2
Approx 1.0% C	—	43.3	43.3	41.5	39.8	36.3	32.9	29.4	27.7	29.4
Approx 1.5% C	—	36.3	36.3	36.3	34.7	32.9	31.2	27.7	27.7	29.4
Tin, Pure	74.4	65.9	58.9	57.1	—	—	—	—	—	—
Tungsten	—	166.2	150.6	141.9	133.4	126.4	112.5	76.2	—	—
Uranium	—	19.18	20.35	23.26	24.19	—	—	—	—	—
Wrought Iron C < 0.5%	—	58.9	57.1	51.9	48.5	45.0	36.3	32.9	32.9	32.9
Zinc, Pure	114.3	112.5	109.1	105.6	100.5	93.5	—	—	—	—
Zirconium	—	—	20.93	20.35	19.89	19.89	21.52	—	—	—
Zirconium 97%, Tin 3%	—	—	11.98	13.26	15.54	15.58	18.02	—	—	—

1 W/mK = 0.86 kcal/m hr °C

VARIATION OF THERMAL CONDUCTIVITY OF METALS WITH TEMPERATURE (W/mK)

Element	−173	−73	127	327	527	727	927	1227	1727
Beryllium	990	301	161	126	106	90.8	78.7	—	—
Bismuth	16.5	9.69	7.04	—	—	—	—	—	—
Boron	190	55.5	16.8	10.6	9.6	9.85	—	—	—
Cadmium	203	99.3	94.7	—	—	—	—	—	—
Chromium	159	111	90.9	80.7	71.3	65.4	61.9	57.2	49.4
Cobalt	167	122	85.4	67.4	58.2	52.1	49.3	42.5	—
Gold	327	323	311	298	284	270	255	—	—
Iridium	172	153	144	138	132	126	120	111	—
Niobium	55.2	52.6	55.2	58.2	61.3	64.4	67.5	72.1	79.1
Palladium	76.5	71.6	73.6	79.7	86.9	94.2	102	110	—
Platinum	77.5	72.6	71.8	73.2	75.6	78.7	82.6	89.3	99.4
Rhodium	186	154	146	136	127	121	116	110	112
Silicon	884	264	98.9	61.9	42.2	31.2	25.7	22.7	—
Tantalum	59.2	57.5	57.8	58.6	59.4	60.2	61.0	62.2	64.1
Titanium	30.5	24.5	20.4	19.4	19.7	20.7	22	24.5	—
Uranium	21.7	25.1	29.6	34.0	38.8	43.9	49.0	—	—
Vanadium	35.8	31.3	31.3	33.3	35.7	38.2	40.8	44.6	50.9
Zirconium	33.2	25.2	21.6	20.7	21.6	23.7	26.0	28.8	33.0

Temperature, °C

1 W/mK = 0.86 kcal/m hr °C

VARIATION OF THERMAL CONDUCTIVITY OF METALS WITH TEMPERATURE

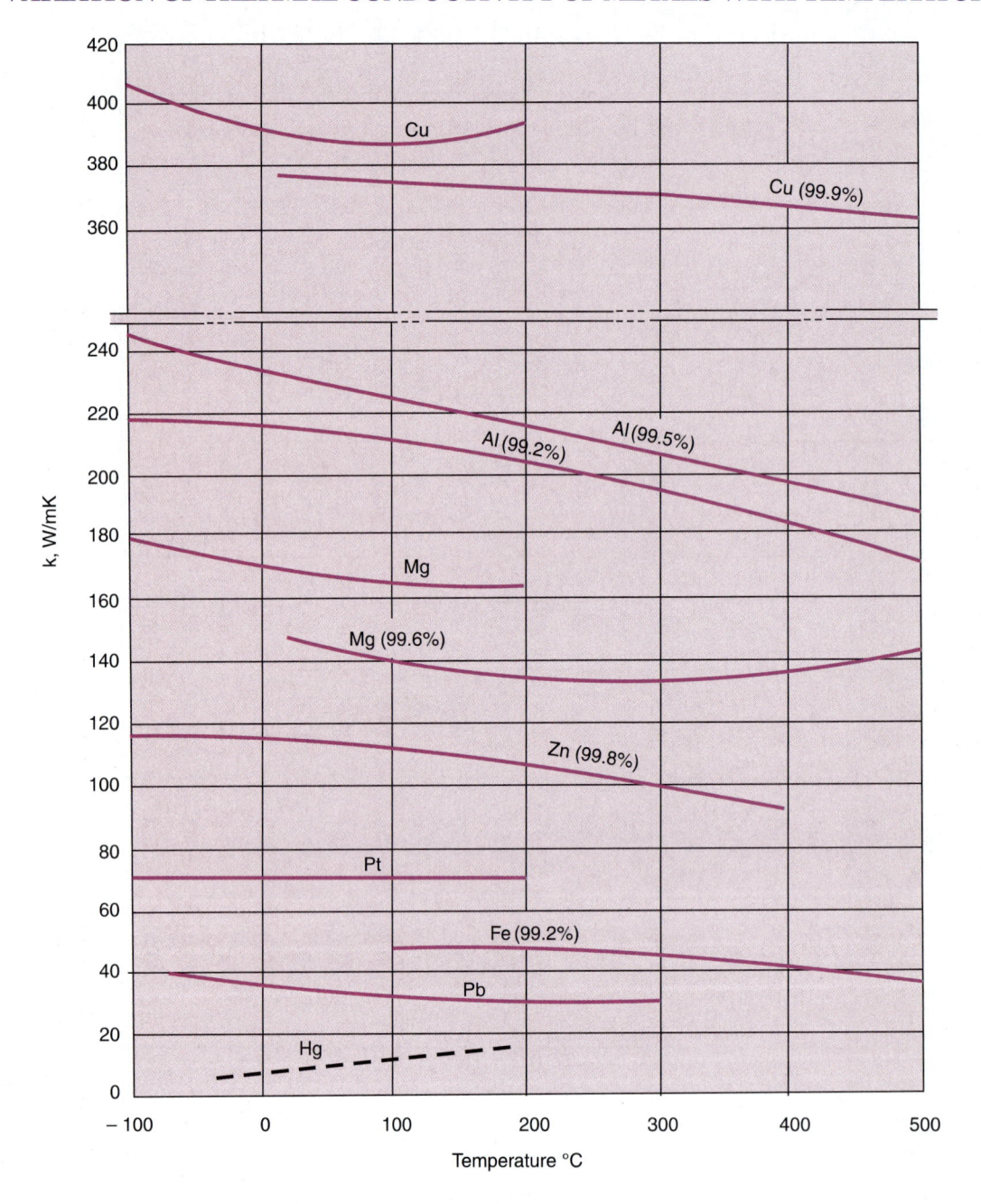

$1\ W/mK = 0.86\ kcal/m\ hr\ °C$

VARIATION OF THERMAL CONDUCTIVITY OF SOME ALLOYS WITH TEMPERATURE (W/mK)

Alloy	-173	-73	127	327	527	727	927	1227
Aluminium alloy 2034 – 76 (4.5% Cu, 1.5% Mg, 0.6% Mn)	65	163	186	186	—	—	—	—
Aluminium alloy 195, Cast, 4.5% Cu	—	—	174	185	—	—	—	—
Commercial Bronze 90% Cu, 10% Al	—	42	52	59	—	—	—	—
Phosphor Bronze (89% Cu, 11% Sn)	—	41	65	74	—	—	—	—
ARMCO Iron	95.6	80.6	65.7	53.1	42.2	32.3	28.7	31.4
Plain Carbon Steel (Mn ≤ 1%, Si ≤ 0.1%)	—	—	56.7	48.0	39.2	30.0	—	—
AISI 110	—	—	58.7	48.8	39.2	31.3	—	—
Carbon Silicon Steel (1% < Mn ≤ 1.65%, 0.1% < Si ≤ 0.6%)	—	—	49.8	44	37.4	29.3	—	—
Carbon-Manganese-Silicon Steel (1% < Mn ≤ 1.65%, 0.1% < Si ≤ 0.6%)	—	—	42.2	39.7	35.0	27.6	—	—
Low Chromium Steel ($\frac{1}{2}$ Cr – $\frac{1}{4}$ Mo – Si) (0.18% C, 0.65% Cr, 0.21% Mo, 0.6% Si)	—	—	38.2	36.7	33.3	26.9	—	—
-do- 1 Cr – $\frac{1}{2}$ Mo (0.16% C, 1% Cr, 0.54% Mo, 0.39% Si)	—	—	42.0	39.1	34.5	27.4	—	—
-do- 1Cr – V (0.2% C, 1.02% Cr, 0.15% V)	—	—	46.8	42.1	36.3	28.2	—	—
Stainless Steels, AISI 302	—	—	17.3	20.0	22.8	25.4	—	—
Stainless Steels AISI 304	9.2	12.6	16.6	19.8	22.6	25.4	28.0	31.7
Stainless Steels AISI 316	—	—	15.2	18.3	21.3	24.2	—	—
Stainless Steels AISI 347	—	—	15.8	18.9	21.9	24.7	—	—
Nicrome (80% Ni, 20% Cr)	—	—	14	16	21	—	—	—
Inconel X-750 (73% Ni, 15% Cr, rest Fe)	8.7	10.3	13.5	17.0	20.5	24.0	27.6	33.0
Platinum Rhodium Alloy (60% Pt, 40% Rh)	—	—	52	59	65	69	73	76

1 W/mK = 0.86 kcal/m hr °C

THERMAL CONDUCTIVITY OF SOME COMMON MATERIALS
AT 25°C, k = W/mK = 0.85984 kcal/hr m °C

Material	k	Material	k
Acetone	0.16	Foam Glass	0.045
Acrylic	0.2	Gasoline	0.15
Alcohol	0.17	Glycerol	0.28
Aluminium 2024, temper 351	143	Hard board–high density	0.15
Aluminium 2024, temper T4	121	Hard woods (oak, maple)	0.16
Aluminium 5052, temper H32	138	Kapak insulation	0.034
Aluminium 5052, temper O	144	Kovar (54% Fe, 29% Ni, 17% Co)	16.3
Aluminium 6061, temper O	144	Lime stone	1.26–1.33
Aluminium 6061, temper T4	154	Magnesite	4.15
Aluminium 6061, temper T6	167	Mercury	8
Aluminium 775, temper T6	128	Methane	0.030
Aluminium A356, temper T6	128	Methanol	0.21
Antimony	18.5	Monel	26
Argon	0.016	Nylon 6	0.25
Asbestos-cement board	0.744	SAE 50, Oil	0.15
Asbestos-cement sheets	0.166	Olive Oil	0.17
Asbestos-cement	2.07	Perlite, Atm. Pressure	0.031
Asbestos loosely packed	0.15	Plywood	0.13
Asbestos mill board	0.14	Polyethylene HD	0.42–0.51
Bitumin	0.17	Polypropylene	0.1–0.22
Benzene	0.18	PTFE	0.25
Brick-dense	1.31	PVC	0.19
Brick work	0.69	Rock solid	2–7
Clay–Dry to Moist	0.15–0.18	Rock porous, volcanic, (Tuff)0	0.5–2.5
Clay saturated	0.6–2.5	Sand stone	1.7
Corian (ceramic filled)	1.06	Silica aerogel	0.02
Cork	0.07	Silicone oil	0.1
Cotton	0.03	Snow (<0°C)	0.05–0.25
Cotton wool insulation	0.029	Solder soft (95% Pb, 5% Sn)	32.3
Diatomaceous earth (Si-o-cel)	0.06	Solder hard (80% Au, 20% Sn)	57
Ether	0.14	Straw insulation	0.09
Epoxy	0.36	Tissue, human skin, at 27°C	0.37
Fiber glass	0.04	Fat layer, at 27°C	0.2
		Muscle, at 27°C	0.41
		Urithane foam	0.021
		Vinyl ether	0.25

PROPERTY VALUES OF SOME NON METALLIC SOLIDS

Material	Temperature °C	Thermal Conductivity k W/mK	Density ρ kg/m^3	Specific Heat c J/kgK
Aluminium oxide, Sapphire	27	46	3970	765
Aluminium oxide, Sapphire	127	32.4	—	940
Aluminium oxide, Sapphire	327	18.9	—	1110
Aluminium oxide, Sapphire	527	13.0	—	1180
Aluminium oxide, Poly crystaline	27	36.0	3970	765
Aluminium oxide, Poly crystaline	127	26.4	—	940
Aluminium oxide, Poly crystaline	327	15.8	—	1110
Aluminium oxide, Poly crystaline	527	10.4	—	1180
Beryllium oxide	27	272.0	3000	1030
Beryllium oxide	127	196.0	—	1350
Beryllium oxide	327	111.0	—	1690
Beryllium oxide	527	70.0	—	1865
Beryllium oxide	727	47.0	—	1975
Boron fibre epoxy composite (30% Vol.)				
Parallel to fibre	27	2.29	2080	1122
Perpendicular to fibre	27	0.59	2080	1122
Carbon, Amorphous	27	1.60	1950	—
Carbon, Amorphous	127	1.89	—	—
Carbon, Amorphous	327	2.19	—	—
Carbon, Amorphous	527	2.37	—	—
Diamond type IIa, insulator	27	2300	3500	509
Diamond type IIa, insulator	127	1540	—	853
Graphite, Pyroletic, parallel to grain	27	1950	2210	709
Graphite, Pyroletic, parallel to grain	127	1390	—	992
Graphite, Pyroletic, parallel to grain	327	892	—	1406
Graphite, Pyroletic, parallel to grain	527	667	—	1650
Graphite, Pyroletic, perpendicular to grain	27	5.70	—	—
Graphite, Pyroletic, perpendicular to grain	127	4.09	—	992
Graphite, Pyroletic, perpendicular to grain	327	2.68	—	1406
Graphite, Pyroletic, perpendicular to grain	527	2.01	—	1650
Graphite fibre Epoxy Composite	27	11.10	1400	935
(25% Volume), parallel to fibre	127	13.00	—	1216
(25% Volume), perpendicular to fibre	27	0.87	—	642
(25% Volume), perpendicular to fibre	127	1.10	—	1216
Pyroceramics, corning 9606	30	3.98	2600	808
Pyroceramics, corning 9606	127	3.64	—	908
Pyroceramics, corning 9606	327	3.28	—	1038
Pyroceramics, corning 9606	527	3.08	—	1122

$1 \, W/mK = 0.86 \, kcal/m \, hr \, °C$ $1 \, J/kgK = 238.9 \times 10^{-6} \, kcal/kg \, °C$ Thermal diffusivity $\alpha = k/\rho c$

PROPERTY VALUES OF SOME NON METALLIC SOLIDS *(Contd.)*

Material	Temperature °C	Thermal Conductivity k W/mK	Density ρ kg/m³	Specific Heat c J/kgK
Silicon carbide	27	490	3160	675
Silicon carbide	727	87	—	1193
Silicon dioxide, Crystalline, (Quartz) Parallel to axis	27	10.4	2650	745
Parallel to axis	127	7.6	—	885
Silicon dioxide, Crystalline (Quartz) Parallel to axis	327	5.0	—	1075
Silicon dioxide, Crystalline, (Quartz) Parallel to axis	527	4.2	—	1250
Silicon dioxide, Crystalline, (Quartz) Perpendicular to grain	27	6.21	—	745
Silicon dioxide, Crystalline, (Quartz) Perpendicular to grain	127	4.70	—	885
Silicon dioxide, Crystalline, (Quartz) Perpendicular to grain	327	3.40	—	1075
Silicon dioxide, Crystalline, (Quartz) Perpendicular to grain	527	3.10	—	1250
Silicon dioxide, Poly crystalline (Fused Silica)	27	1.38	2220	745
(Fused Silica)	127	1.51	—	905
(Fused Silica)	327	1.75	—	1040
(Fused Silica)	527	2.17	—	1105
Silicon Nitride	27	16	2400	691
Silicon Nitride	127	13.9	—	778
Silicon Nitride	327	11.3	—	937
Silicon Nitride	527	9.88	—	1063
Sulphur	27	0.206	2070	708
Sulphur	−73	0.185	—	606
Thorium dioxide	27	13	9110	235
Thorium dioxide	127	10.2	—	255
Thorium dioxide	327	6.6	—	274
Thorium dioxide	527	4.7	—	285
Thorium dioxide	727	3.68	—	293
Titanium dioxide	27	8.4	4157	710
Poly Crystalline	127	7.01	—	805
Poly Crystalline	327	5.02	—	880
Poly Crystalline	527	3.94	—	910
Poly Crystalline	727	3.46	—	930

1 W/mK = 0.86 kcal/m hr °C   1 J/kg K = 238.9 × 10^{-6} kcal/kg °C   Thermal diffusivity $\alpha = k/\rho c$

HEAT AND MASS TRANSFER DATA BOOK

PROPERTY VALUES OF COMMON MATERIALS

Material	Temperature °C	Thermal Conductivity k W/mK	Density ρ kg/m³	Specific Heat c J/kgK
Asphalt	27	0.062	2115	920
Bakelite	27	1.40	1300	1465
Refractory carborundum brick	599	18.5	—	—
Refractory carborundum brick	1399	11.0	—	—
Chrome brick	200	2.3	3010	835
Chrome brick	550	2.5	—	—
Chrome brick	900	2.0	—	—
Diatomaceous silica brick, fired	205	0.25	—	—
Diatomaceous silica brick, fired	872	0.30	—	—
Fire clay brick burnt at 1327°C	500	1.00	2050	960
Fire clay brick burnt at 1327°C	800	1.10	—	—
Fire clay brick burnt at 1327°C	1100	1.10	—	—
Fire clay brick burnt at 1452°C	500	1.30	2325	960
Fire clay brick burnt at 1452°C	800	1.40	—	—
Fire clay brick burnt at 1452°C	1100	1.40	—	—
Fire clay brick	205	1.00	2645	960
Fire clay brick	649	1.50	—	—
Fire clay brick	1205	1.80	—	—
Magnesite brick	205	3.8	—	1130
Magnesite brick	649	2.8	—	—
Magnesite brick	1205	1.9	—	—
Clay	27	1.3	1460	880
Coal, anthracite	27	0.26	1350	1260
Concrete (Stone mix)	27	1.40	2300	880
Cotton	27	0.06	80	—
Food materials. Banana (75.5% water content)	27	0.481	980	3350
Apple (75% water content)	27	0.513	840	3600
Cake batter	27	0.223	720	—
Fully baked cake	27	0.121	280	—
Chicken meat, white (74.4% water content)	−75	1.60	—	—
	−40	1.49	—	—
Chicken meat, white (74.4% water content)	−20	1.35	—	—
Chicken meat, white (74.4% water content)	−10	1.20	—	—
Chicken meat, white (74.4% water content)	0	0.476	—	—

1 W/mK = 0.86 kcal/m hr °C   Thermal diffusivity $\alpha = k/\rho c$ 1 J/kg K = 238.9 × 10⁻⁶ kcal/kg °C

PROPERTY VALUES OF COMMON MATERIALS (Contd.)

Material	Temperature °C	Thermal Conductivity k W/mK	Density ρ kg/m³	Specific Heat c J/kgK
Chicken meat, white (74.4% water content)	10	0.48	—	—
Chicken meat, white (74.4% water content)	20	0.489	—	—
Glass, plate (Soda lime)	27	1.4	2500	750
Pyrex glass	27	1.4	2225	835
Ice	0	1.88	920	2040
Ice	−20	2.03	—	1945
Leather sole	27	0.159	998	—
Paper	27	0.180	930	1340
Paraffin	27	0.240	900	2890
Granite	27	2.790	2630	775
Limestone	27	2.150	2320	810
Marble	27	2.80	2680	830
Quartsite	27	5.38	2640	1105
Sand Stone	27	2.90	2150	745
Vulcanized rubber, soft	27	0.13	1100	2010
Vulcanized rubber, hard	27	0.16	1190	—
Sand	27	0.27	1515	800
Soil	27	0.52	2050	1840
Snow	0	0.049	110	—
Snow compacted	0	0.190	500	—
Teflon	27	0.35	2200	—
Teflon	127	0.45	—	—
Human skin	27	0.37	—	—
Human fat layer	27	0.20	—	—
Human Muscle	27	0.41	—	—
Wood, cross grain				
Wood, Balso	27	0.055	140	—
Wood Cypruss	27	0.097	465	—
Wood Fir	27	0.11	415	2720
Wood Oak	27	0.17	545	2385
Wood yellow pine	27	0.15	640	2805
Wood white pine	27	0.11	435	—
Wood radial				
Wood oak	27	0.19	545	2385
Wood fir	27	0.14	420	2720

1 W/mK = 0.86 kcal/m hr °C   Thermal diffusivity $\alpha = k/\rho c$ 1 J/kg K = 238.9 × 10⁻⁶ kcal/kg °C

PROPERTY VALUES OF INSULATING MATERIALS AND SYSTEMS AT 27°C

Material	Density ρ kg/m³	Thermal Conductivity k W/mK	Specific Heat c J/kgK
Blanket and Batt			
Glass fibre, paper based	16	0.046	—
Glass fibre, paper based	28	0.038	—
Glass fibre, paper based	40	0.035	—
Glass fibre, coated, duct liner	32	0.038	835
Board and slab			
Cellular glass	145	0.058	1000
Glass fibre organic bonded	105	0.036	795
Polystyrene, expanded, extruded	55	0.027	1210
Polystyrene Moulded Beads	16	0.040	1210
Mineral fibre board, roofing material	265	0.049	—
Wood, shredded and Cemented	350	0.087	1590
Cork	120	0.039	1800
Loose fill			
Cork, granulated	160	0.045	—
Diatomaceous silica, Coarse	350	0.069	—
Diatomaceous silica, Powder	400	0.091	—
Glass fibre, poured or blown	16	0.043	835
Vermiculate, flakes	80	0.068	835
Vermiculate, flakes	160	0.063	1000
Formed/Foamed in place			
Mineral wool granules with asbestos inorganic binders, sprayed	190	0.046	—
Polyvinyl acetate cork mastic			
Sprayed and troweled	—	0.100	—
Urethane, two part mixture rigid foam	70	0.026	1045
Reflective :			
Aluminium foil Separating fluffy Glass mats ; 10 to 12 layers, evacuated For cryogenic applications (150 K)	40	160×10^{-6}	—
Aluminium foil and glass paper Laminate 75-150 layers, evacuated For cryogenic applications (150 k)	120	17×10^{-6}	—
Typical silica powder, evacuated	160	1.7×10^{-3}	—

1 W/mK = 0.86 kcal/m hr °C

1 J/kg K = 238.9 $\times 10^{-6}$ kcal/kg °C

PROPERTY VALUES OF INDUSTRIAL INSULATING MATERIALS

Material	Density ρ kg/m³	Temperature °C	Thermal Conductivity k W/mK
Evacuated super insulation	Variable	−240 to 1000	$1.5 \times 10^{-6} - 720 \times 10^{-6}$
BLANKETS :			
Alumina silica fibre blanket	48	255	0.071
Alumina silica fibre blanket	48	370	0.150
Alumina silica fibre blanket	64	255	0.059
Alumina silica fibre blanket	64	370	0.087
Alumina silica fibre blanket	96	255	0.052
Alumina silica fibre blanket	96	370	0.076
Alumina silica fibre blanket	128	255	0.049
Alumina silica fibre blanket	128	370	0.068
Elastemeric Sheets	70–100	−40–100	0.036–0.039
Felt blanket, semi rigid	50–25	−3	0.035
Felt blanket, semi rigid	50–25	27	0.038
Felt blanket, semi rigid	50–25	92	0.051
Felt blanket, organic bonded	50	−3	0.030
Felt blanket, organic bonded	50	27	0.038
Felt blanket, organic bonded	50	92	0.051
Felt blanket, laminated without binder	120	147	0.051
Felt blanket, laminated without binder	120	257	0.065
Felt blanket, laminated without binder	120	372	0.087
Fibre glass blanket, organic bonded	12	27	0.046
Fibre glass blanket, organic bonded	12	92	0.069
Fibre glass blanket, organic bonded	16	27	0.042
Fibre glass blanket, organic bonded	16	92	0.062
Fibre glass blanket, organic bonded	24	27	0.039
Fibre glass blanket, organic bonded	24	92	0.053
Fibre glass blanket, organic bonded	32	27	0.033
Fibre glass blanket, organic bonded	32	92	0.048
Fibre glass blanket, organic bonded	48	27	0.033
Fibre glass blanket, organic bonded	48	92	0.045
Fibre glass blanket with vapour barrier	5–70	10–32	0.029–0.045
Mineral fibre blanket	10	−3	0.040
Mineral fibre blanket	10	12	0.043

1 W/mK = 0.86 kcal/m hr °C

PROPERTY VALUES OF INDUSTRIAL INSULATING MATERIALS *(Contd.)*

Material	Density ρ kg/m^3	Temperature °C	Thermal Conductivity k W/mK
Mineral fibre blanket	10	27	0.048
Mineral fibre blanket	10	37	0.052
Mineral fibre blanket	90–192	37	0.038
Mineral fibre blanket	90–192	92	0.046
Mineral fibre blanket	90–192	147	0.056
Mineral fibre blanket, metal reinforced	40–96	37	0.035
Mineral fibre blanket, metal reinforced	40–96	92	0.045
Mineral fibre blanket, metal reinforced	40–96	147	0.058
Mineral fibre blanket, metal reinforced	40–96	257	0.088
BLOCKS ; Board and Pipe Insulation :			
Asbestos paper laminated and corrugated 4 ply	190	27	0.078
Asbestos paper laminated and corrugated 4 ply	190	92	0.098
Asbestos paper laminated and corrugated 6 ply	255	27	0.071
Asbestos paper laminated and corrugated 6 ply	255	92	0.085
Asbestos paper laminated and corrugated 8 ply	300	27	0.068
Asbestos paper laminated and corrugated 8 ply	300	92	0.082
Calcium Silicate Blocks	190	37	0.055
Calcium Silicate Blocks	190	92	0.059
Calcium Silicate Blocks	190	147	0.063
Calcium Silicate Blocks	190	257	0.075
Cellular glass blocks	145	−33	0.048
Cellular glass blocks	145	−3	0.052
Cellular glass blocks	145	27	0.058
Cellular glass blocks	145	92	0.068
Diatomaceous Silica blocks	345	257	0.092
Diatomaceous Silica blocks	345	372	0.098
Diatomaceous Silica blocks	385	257	0.101
Diatomaceous Silica blocks	385	372	0.100
Insulating Cement-mineral fibre	430	37	0.071
– with clay binder	430	92	0.074
– with clay binder	430	147	0.088

1 W/mK = 0.86 kcal/m hr °C

PROPERTY VALUES OF INDUSTRIAL INSULATING MATERIALS *(Contd.)*

Material	Density ρ kg/m³	Temperature °C	Thermal Conductivity k W/mK
Insulating Cement-mineral with hydraulic binder	560	37	0.108
Insulating Cement-mineral with hydraulic binder	560	92	0.115
Insulating Cement-mineral with hydraulic binder	560	147	0.123
Magnesia, 85% blocks	185	37	0.051
Magnesia, 85% blocks	185	92	0.058
Magnesia, 85% blocks	185	147	0.061
Mineral fibre preformed blocks	125–160	upto 650	0.035–0.091
Polystyrene, rigid, extruded	56	–73	0.023
Polystyrene, rigid, extruded	56	–3	0.025
Polystyrene, rigid, extruded	56	27	0.027
Polystyrene, rigid, extruded	35	–73	0.023
Polystyrene, rigid, extruded	35	–3	0.026
Polystyrene, rigid, extruded	35	27	0.029
Polystyrene, Moulded beads	16	–73	0.026
Polystyrene, Moulded beads	16	–3	0.036
Polystyrene, Moulded beads	16	27	0.040
Rubber, rigid formed	70	–3	0.029
Rubber, rigid formed	70	27	0.032
Urethane foam blocks	25–65	100–150	0.016–0.020
Loose Fill :			
Cellulose, wood or paper pulp	45	27	0.039
Cellulose, wood or paper pulp	45	37	0.042
Perlite, expanded	105	–73	0.036
Perlite, expanded	105	–33	0.043
Perlite, expanded	105	–3	0.49
Perlite, expanded	105	27	0.053
Vermiculite expanded	122	–43	0.056
Vermiculite expanded	122	–3	0.063
Vermiculite expanded	122	–27	0.068
Vermiculite expanded	80	–43	0.049
Vermiculite expanded	80	–3	0.058
Vermiculite expanded	80	–27	0.063

1 W/mK = 0.86 kcal/m hr °C

PROPERTY VALUES OF INSULATING, BUILDING AND OTHER MATERIALS

Material	Temperature t °C	Thermal Conductivity k W/mK	Density ρ kg/m³	Specific Heat c J/kgK
Aluminium foil	50	0.0465	20	—
Asbestos, fibre	50	0.1105	470	816
Asphalt	20	0.6978	2110	2093
Balsa wood	30	0.0523	128	—
Boiler Scale	65	1.314–3.1400	—	—
Building brick	20	0.2330–0.2910	800–1500	—
Chrome brick	200	2.320	3000	840
Chrome brick	550	2.470	3000	840
Chrome brick	900	1.990	3000	840
Corrugated card board	20	0.0640	—	—
Cellotex	20	0.0465	215	—
Celluloid	30	0.2093	1400	—
Chalk	50	0.9304	2000	879
Clinker	30	0.1628	1400	1675
Coal	20	0.1861	1400	1306
Coke, powdered	100	0.1907	419	1214
Concrete	20	1.2790	2300	1130
Cork, granulated	20	0.0384	45	—
Cork, plate	30	0.0419	190	1884
Earth, dry	—	0.1384	1500	—
Earth, wet	—	0.6571	1700	2010
Fibre plate	20	0.0489	240	—
Fibre brick	100	0.1396	550	—
Glass	20	0.7443	2500	670
Glass wool	20	0.0372	200	670
Gravel	20	0.3605	1840	—
Granite	20	0.00279	2630	775
Gypsum	—	0.2908	1650	—
Ice	0	2.2500	920	2261
Ice	−95	3.9540	—	1172
Lamp black	40	0.1477	190	—
Leather Sole	30	0.1593	1000	—
Linoleum	20	0.1861	1180	—
Magnesia, 85% powdered	100	0.0675	216	—

1 W/mk = 0.86 kcal/m hr°C   *1 J/kg K = 238.9 × 10⁻⁶ kcal/kg°C*   *Thermal diffusivity, $\alpha = k/\rho c$.*

PROPERTY VALUES OF INSULATING, BUILDING AND OTHER MATERIALS
(Contd.)

Material	Temperature t °C	Thermal Conductivity k W/mK	Density ρ kg/m^3	Specific Heat c J/kgK
Marble	90	2.9400	2700	800
Mica	—	0.5815	290	879
Mineral wool	50	0.0465	200	921
Oak, across grain	20	0.2070	800	1759
Oak, along grain	20	0.3629	800	—
Paraffin	20	0.2675	920	—
Peat, plate	50	0.0640	220	—
Pine, across grain	20	0.1070	448	—
Pine, along grain	20	0.2559	448	—
Plaster	20	0.7792	1680	—
Porcelain	95	1.0350	2400	1089
Porcelain	1065	1.9650	2400	—
Polystyrene	20	0.1570	1050	—
Portland cement	30	0.3024	1900	1130
Quartz, across grain	0	7.2110	2500–2800	837
Quartz, along grain	0	13.607	2500–2800	837
Refractory clay	450	1.036	1845	1089
Rubber	0	0.1628	1200	1382
Sand, dry	20	0.3256	1500	796
Sand, damp	20	1.128	1650	2093
Saw dust	20	0.698	200	—
Sheet, Asbestos	30	0.1163	770	816
Slag Concrete. lumps	—	0.9304	2150	879
Slag wool	100	0.6980	250	—
Slate	100	1.4890	2800	—
Snow	—	0.4652	560	2093
Sugar, granulated	0	0.5815	1600	1256
Wool, felt	30	0.0523	330	—

1 W/mK = 0.86 kcal/m hr °C   1 J/kg K = 238.9 × 10^{-6} kcal/kg °C   Thermal diffusivity, $\alpha = k/\rho c$.

HEAT AND MASS TRANSFER DATA BOOK

VARIATION OF THERMAL CONDUCTIVITY OF DIFFERENT INSULATING AND REFRACTORY MATERIALS WITH TEMPERATURE

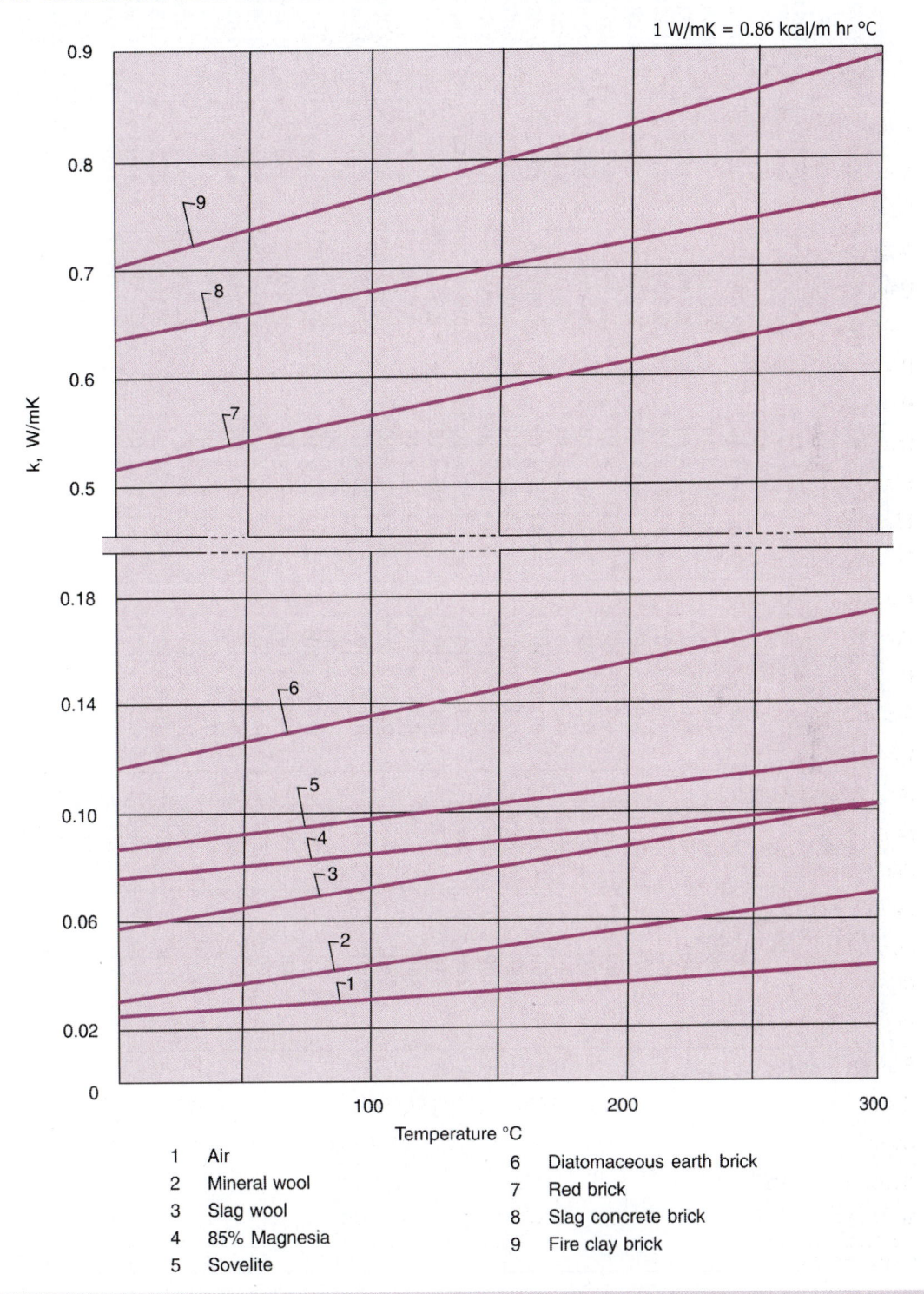

1 W/mK = 0.86 kcal/m hr °C

1 Air	6 Diatomaceous earth brick
2 Mineral wool	7 Red brick
3 Slag wool	8 Slag concrete brick
4 85% Magnesia	9 Fire clay brick
5 Sovelite	

PROPERTY VALUES OF LIQUIDS IN SATURATED STATE

	Temperature t °C	Density ρ kg/m³	Kinematic Viscosity ν m²/s	Thermal Diffusivity α m²/s	Prandtl Number Pr	Specific Heat c J/kgK	Thermal Conductivity k W/mK
WATER	0	1002	1.788×10^{-6}	0.1308×10^{-6}	13.600	4216	0.5524
	20	1000	1.006×10^{-6}	0.1431×10^{-6}	7.020	4178	0.5978
	40	995	0.657×10^{-6}	0.1511×10^{-6}	4.340	4178	0.6280
	60	985	0.478×10^{-6}	0.1553×10^{-6}	3.020	4183	0.6513
	80	974	0.364×10^{-6}	0.1636×10^{-6}	2.220	4195	0.6687
	100	961	0.293×10^{-6}	0.1681×10^{-6}	1.740	4216	0.6804
	120	945	0.247×10^{-6}	0.1708×10^{-6}	1.446	4250	0.6850
	140	928	0.213×10^{-6}	0.1725×10^{-6}	1.241	4283	0.6838
	160	909	0.189×10^{-6}	0.1728×10^{-6}	1.099	4342	0.6804
	180	889	0.173×10^{-6}	0.1725×10^{-6}	1.044	4417	0.6757
	200	867	0.160×10^{-6}	0.1701×10^{-6}	0.937	4505	0.6652
	220	842	0.149×10^{-6}	0.1681×10^{-6}	0.891	4610	0.6524
	240	815	0.143×10^{-6}	0.1639×10^{-6}	0.871	4756	0.6350
	260	786	0.137×10^{-6}	0.1578×10^{-6}	0.874	4949	0.6106
	280	752	0.135×10^{-6}	0.1481×10^{-6}	0.910	5208	0.5803
	300	714	0.135×10^{-6}	0.1325×10^{-6}	1.019	5728	0.5396
AMMONIA	-50	704	0.435×10^{-6}	0.1742×10^{-6}	2.600	4463	0.5466
	-40	691	0.406×10^{-6}	0.1775×10^{-6}	2.280	4467	0.5466
	-30	679	0.387×10^{-6}	0.1800×10^{-6}	2.150	4476	0.5489
	-20	667	0.381×10^{-6}	0.1819×10^{-6}	2.090	4509	0.5466
	-10	653	0.378×10^{-6}	0.1825×10^{-6}	2.070	4564	0.5431
	0	640	0.373×10^{-6}	0.1819×10^{-6}	2.050	4635	0.5396
	10	626	0.368×10^{-6}	0.1800×10^{-6}	2.040	4714	0.5315
	20	612	0.358×10^{-6}	0.1775×10^{-6}	2.020	4798	0.5210
	30	596	0.350×10^{-6}	0.1742×10^{-6}	2.010	4890	0.5071
	40	581	0.340×10^{-6}	0.1700×10^{-6}	2.000	4999	0.4931
	50	564	0.330×10^{-6}	0.1656×10^{-6}	1.990	5116	0.4757

For values of β refer page 30 or $\beta = (\Delta \rho)/(\rho \times \Delta T)$ (use $\Delta \rho$, ρ, ΔT from tables), $\mu = \rho \nu$, $1 W/mK = 0.86 \, kcal/m \, hr \, °C$, $1 J/kg \, K = 238.9 \times 10^{-6} \, kcal/kg \, °C$

PROPERTY VALUES OF LIQUIDS IN SATURATED STATE (Contd.)

	Temperature t °C	Density ρ kg/m^3	Kinematic Viscosity v m^2/s	Thermal Diffusivity α m^2/s	Prandtl Number Pr	Specific Heat c J/kgK	Thermal Conductivity k W/mK
CARBON-DIOXIDE	−50	1156	0.119×10^{-6}	0.0403×10^{-6}	2.96	1842	0.0855
	−40	1118	0.118×10^{-6}	0.0481×10^{-6}	2.46	1884	0.1011
	−30	1077	0.117×10^{-6}	0.0528×10^{-6}	2.22	1968	0.1117
	−20	1032	0.115×10^{-6}	0.0544×10^{-6}	2.12	2652	0.1150
	−10	983	0.113×10^{-6}	0.0514×10^{-6}	2.22	2177	0.1099
	0	927	0.108×10^{-6}	0.0458×10^{-6}	2.38	2470	0.1046
	10	863	0.102×10^{-6}	0.0361×10^{-6}	2.80	3140	0.0971
	20	772	0.091×10^{-6}	0.0222×10^{-6}	4.10	5024	0.0872
	30	598	0.079×10^{-6}	0.0028×10^{-6}	28.70	36425	0.0703
GLYCERINE-$C_3H_5(OH)_3$	0	1276	8314×10^{-6}	0.0983×10^{-6}	84700	2261	0.2826
	10	1270	3000×10^{-6}	0.0964×10^{-6}	31000	2320	0.2838
	20	1264	1180×10^{-6}	0.0947×10^{-6}	12500	2387	0.2861
	30	1258	501×10^{-6}	0.0928×10^{-6}	5380	2445	0.2861
	40	1252	223×10^{-6}	0.0914×10^{-6}	2450	2512	0.2861
	50	1245	149×10^{-6}	0.0892×10^{-6}	1630	2583	0.2873
SULPHUR DIOXIDE	−50	1561	0.484×10^{-6}	0.1142×10^{-6}	4.24	1361	0.2419
	−40	1537	0.423×10^{-6}	0.1131×10^{-6}	3.74	1361	0.2349
	−30	1521	0.370×10^{-6}	0.1119×10^{-6}	3.31	1361	0.2303
	−20	1488	0.324×10^{-6}	0.1108×10^{-6}	2.93	1361	0.2256
	−10	1463	0.288×10^{-6}	0.1197×10^{-6}	2.62	1361	0.2186
	0	1438	0.257×10^{-6}	0.1081×10^{-6}	2.38	1365	0.2117
	10	1412	0.232×10^{-6}	0.1067×10^{-6}	2.18	1365	0.2047
	20	1386	0.209×10^{-6}	0.1050×10^{-6}	2.00	1365	0.1989
	30	1359	0.189×10^{-6}	0.1036×10^{-6}	1.83	1365	0.1919
	40	1329	0.173×10^{-6}	0.1019×10^{-6}	1.70	1369	0.1849
	50	1299	0.162×10^{-6}	0.1000×10^{-6}	1.61	1369	0.1768

For the values of β refer page 30 or $\beta = (\Delta \rho)/(\rho \times \Delta T)$ (use $\Delta \rho$, ρ, ΔT from tables), $\mu = \rho v$, $1 \, W/mK = 0.86 \, kcal/m \, hr \, °C$, $1 \, J/kg \, K = 238.9 \times 10^{-6} \, kcal/kg \, °C$

PROPERTY VALUES OF LIQUIDS IN SATURATED STATE (Contd.)

Temperature t °C	Density ρ kg/m³	Kinematic Viscosity ν m²/s	Thermal Diffusivity α m²/s	Prandtl Number Pr	Specific Heat c J/kgK	Thermal Conductivity k W/mK
METHYL CHLORIDE (REFRIGERANT-40)-CH$_3$Cl						
-50	1052	0.319 × 10⁻⁶	0.1389 × 10⁻⁶	2.31	1474	0.2512
-40	1033	0.317 × 10⁻⁶	0.1367 × 10⁻⁶	2.32	1482	0.2093
-30	1016	0.314 × 10⁻⁶	0.1336 × 10⁻⁶	2.35	1491	0.2024
-20	999	0.309 × 10⁻⁶	0.1300 × 10⁻⁶	2.38	1503	0.1954
-10	981	0.305 × 10⁻⁶	0.1256 × 10⁻⁶	2.43	1520	0.1872
0	962	0.302 × 10⁻⁶	0.1214 × 10⁻⁶	2.49	1537	0.1779
10	942	0.293 × 10⁻⁶	0.1167 × 10⁻⁶	2.55	1562	0.1710
20	923	0.292 × 10⁻⁶	0.1111 × 10⁻⁶	2.63	1587	0.1628
30	903	0.288 × 10⁻⁶	0.1058 × 10⁻⁶	2.72	1616	0.1535
40	883	0.281 × 10⁻⁶	0.0997 × 10⁻⁶	2.83	1650	0.1442
50	861	0.274 × 10⁻⁶	0.0925 × 10⁻⁶	2.97	1687	0.1338
DICHLORO DIFLURO METHANE (REFRIGERANT-12)-CCl$_2$F$_2$						
-50	1547	0.310 × 10⁻⁶	0.0500 × 10⁻⁶	6.20	875	0.0675
-40	1518	0.279 × 10⁻⁶	0.0514 × 10⁻⁶	5.40	883	0.0692
-30	1490	0.253 × 10⁻⁶	0.0528 × 10⁻⁶	4.80	896	0.0692
-20	1460	0.235 × 10⁻⁶	0.0539 × 10⁻⁶	4.40	909	0.0709
-10	1429	0.220 × 10⁻⁶	0.0550 × 10⁻⁶	4.00	917	0.0727
0	1397	0.213 × 10⁻⁶	0.0558 × 10⁻⁶	3.80	934	0.0727
10	1364	0.203 × 10⁻⁶	0.0561 × 10⁻⁶	3.60	950	0.0727
20	1330	0.198 × 10⁻⁶	0.0561 × 10⁻⁶	3.50	963	0.0727
30	1295	0.194 × 10⁻⁶	0.0561 × 10⁻⁶	3.50	984	0.0709
40	1257	0.191 × 10⁻⁶	0.0556 × 10⁻⁶	3.50	1005	0.0692
50	1216	0.189 × 10⁻⁶	0.0544 × 10⁻⁶	3.50	1022	0.0672
EUTECTIC CALCIUM CHLORIDE SOLUTION (29.9% CaCl$_2$)						
-50	1320	36.352 × 10⁻⁶	0.1167 × 10⁻⁶	312	2608	0.4012
-40	1314	24.971 × 10⁻⁶	0.1200 × 10⁻⁶	208	2634	0.4152
-30	1310	17.177 × 10⁻⁶	0.1233 × 10⁻⁶	139	2663	0.4292

For values of β refer page 30 or $\beta = (\Delta \rho)/(\rho \times \Delta T)$ (use $\Delta \rho$, ρ, ΔT from tables), $\mu = \rho \nu$, $1\ W/mK = 0.86\ kcal/m\ hr\ °C$, $1\ J/kg\ K = 238.9 \times 10^{-6}\ kcal/kg\ °C$

HEAT AND MASS TRANSFER DATA BOOK

PROPERTY VALUES OF LIQUIDS IN SATURATED STATE (Contd.)

Temperature t °C	Density ρ kg/m³	Kinematic Viscosity ν m²/s	Thermal Diffusivity α m²/s	Prandtl Number Pr	Specific Heat c J/kg K	Thermal Conductivity k W/mK
29.9% CaCl$_2$ solution (Contd.)						
−20	1305	11.036 × 10⁻⁶	0.1267 × 10⁻⁶	87.10	2688	0.4454
−10	1300	6.956 × 10⁻⁶	0.1300 × 10⁻⁶	53.60	2713	0.4582
0	1296	4.394 × 10⁻⁶	0.1331 × 10⁻⁶	33.00	2738	0.4722
10	1291	3.353 × 10⁻⁶	0.1364 × 10⁻⁶	24.60	2763	0.4850
20	1286	2.722 × 10⁻⁶	0.1394 × 10⁻⁶	19.60	2788	0.4989
30	1282	2.267 × 10⁻⁶	0.1422 × 10⁻⁶	16.00	2813	0.5106
40	1277	1.923 × 10⁻⁶	0.1444 × 10⁻⁶	13.30	2839	0.5234
50	1272	1.654 × 10⁻⁶	0.1469 × 10⁻⁶	11.30	2868	0.5350
ETHYLENE GLYCOL — C$_2$H$_4$(OH)$_2$						
0	1131	57.530 × 10⁻⁶	0.0933 × 10⁻⁶	615.00	2294	0.2419
20	1116	19.174 × 10⁻⁶	0.0939 × 10⁻⁶	204.00	2382	0.2489
40	1101	8.686 × 10⁻⁶	0.0939 × 10⁻⁶	93.00	2474	0.2559
60	1087	4.747 × 10⁻⁶	0.0931 × 10⁻⁶	51.00	2562	0.2594
80	1077	2.982 × 10⁻⁶	0.0922 × 10⁻⁶	32.40	2650	0.2617
100	1058	2.025 × 10⁻⁶	0.0908 × 10⁻⁶	22.40	2742	0.2628
ENGINE OIL (UNUSED)						
0	899	4282 × 10⁻⁶	0.0911 × 10⁻⁶	47100.00	1796	0.1477
20	888	901 × 10⁻⁶	0.0872 × 10⁻⁶	10400.00	1880	0.1454
40	876	241 × 10⁻⁶	0.0833 × 10⁻⁶	2870.00	1964	0.1442
60	864	83 × 10⁻⁶	0.0800 × 10⁻⁶	1050.00	2047	0.1407
80	852	37 × 10⁻⁶	0.0769 × 10⁻⁶	490.00	2131	0.1384
100	840	20 × 10⁻⁶	0.0739 × 10⁻⁶	276.00	2219	0.1372
120	828	12 × 10⁻⁶	0.0711 × 10⁻⁶	175.00	2307	0.1349
140	816	8 × 10⁻⁶	0.0686 × 10⁻⁶	116.00	2395	0.1338
160	805	5 × 10⁻⁶	0.0664 × 10⁻⁶	84.00	2483	0.1314

For values of β refer page 30 or $\beta = (\Delta\rho)/(\rho \times \Delta T)$ (use $\Delta\rho$, ρ, ΔT from tables), $\mu = \rho\nu$, $1\ W/mK = 0.86\ kcal/m\ hr\ °C$, $1\ J/kgK = 238.9 \times 10^{-6}\ kcal/kg\ °C$

PROPERTY VALUES OF TRANSFORMER OIL (STANDARD 982—68)

Temp. t °C	Density ρ kg/m³	Kinematic Viscosity ν m²/s	Thermal Diffusivity α m²/s	Prandtl Number Pr	Specific Heat c J/kgK	Thermal Conductivity k W/mK	Absolute Viscosity μ Ns/m²
−50	922	$31,800 \times 10^{-6}$	74.2×10^{-9}	42860×10^{-3}	1700	0.116	$29,320 \times 10^{-3}$
−40	916	$4,220 \times 10^{-6}$	75.0×10^{-9}	5630×10^{-3}	1680	0.116	$3,866 \times 10^{-3}$
−30	910	$1,300 \times 10^{-6}$	76.4×10^{-9}	1700×10^{-3}	1650	0.115	$1,183 \times 10^{-3}$
−20	904	404×10^{-6}	77.8×10^{-9}	520×10^{-3}	1620	0.114	365.6×10^{-3}
−10	898	120×10^{-6}	78.8×10^{-9}	153×10^{-3}	2788	0.113	108.1×10^{-3}
0	891	67.5×10^{-6}	77.8×10^{-9}	86.7×10^{-3}	2813	0.112	55.24×10^{-3}
10	885	37.8×10^{-6}	76.3×10^{-9}	49.5×10^{-3}	2839	0.111	33.45×10^{-3}
20	879	24.0×10^{-6}	73.6×10^{-9}	32.6×10^{-3}	2868	0.111	21.10×10^{-3}
30	873	15.4×10^{-6}	70.7×10^{-9}	21.8×10^{-3}	2382	0.110	13.44×10^{-3}
40	867	10.8×10^{-6}	68.8×10^{-9}	15.7×10^{-3}	2474	0.109	9.364×10^{-3}

1 W/mK = 0.86 kcal/m hr °C   1 J/kg K = 238.9×10^{-6} kcal/kg °C   1 Ns/m² = 0.102 kgfs/m²

PROPERTY VALUES OF MOBILTHERM 600

Temp. t °C	Density ρ kg/m³	Coefficient of Thermal Expansion β 1/K	Kinematic Viscosity ν m²/s	Thermal Diffusivity α m²/s	Prandtl Number Pr	Specific Heat c J/kg K	Thermal Conductivity k W/mK	Absolute Viscosity μ Ns/m²
10	953	0.621×10^{-3}		83.3×10^{-9}		1549	0.123	
50	929	0.637×10^{-3}	32.60×10^{-6}	76.9×10^{-9}	424	1680	0.120	30.28×10^{-3}
100	899	0.658×10^{-3}	6.10×10^{-6}	69.4×10^{-9}	87.9	1859	0.116	5.48×10^{-3}
150	870	0.680×10^{-3}	2.34×10^{-6}	64.0×10^{-9}	36.6	2031	0.113	2.04×10^{-3}
200	839	0.705×10^{-3}	1.25×10^{-6}	59.4×10^{-9}	21.0	2209	0.110	1.05×10^{-3}
250	810	0.730×10^{-3}	0.790×10^{-6}	54.5×10^{-9}	14.5	2386	0.106	0.64×10^{-3}

1 W/mK = 0.86 kcal/m hr °C   1 J/kg K = 238.9×10^{-6} kcal/kg °C   1 Ns/m² = 0.102 kgfs/m²

PROPERTY VALUES OF n-BUTYL ALCOHOL ($C_4H_{10}O$)

Temp. t °C	Density ρ kg/m³	Coefficient of Thermal Expansion β 1/K	Kinematic Viscosity ν m²/s	Thermal Diffusivity α m²/s	Prandtl Number Pr	Specific Heat c J/kgK	Thermal Conductivity k W/mK	Absolute Viscosity μ Ns/m²
16	809		4.16×10^{-6}	159.1×10^{-9}	0.261	1305	0.168	3.36×10^{-3}
38	796	81×10^{-3}	2.41×10^{-6}	149.8×10^{-9}	0.161	1392	0.166	1.92×10^{-3}
66	777	86×10^{-3}	1.29×10^{-6}	140.5×10^{-9}	0.0916	1502	0.164	1.00×10^{-3}
93	756		0.76×10^{-6}	134.0×10^{-9}	0.056	1609	0.163	0.57×10^{-3}
117.5	737		0.53×10^{-6}	129.6×10^{-9}	0.041	1706	0.163	0.39×10^{-3}
149						1830		0.28×10^{-3}

1 W/mK = 0.86 kcal/m hr °C   1 J/kg K = 238.9 × 10⁻⁶ kcal/kg °C   1 Ns/m² = 0.102 kgfs/m²

PROPERTY VALUES OF MOLTEN SALT (EQUIMOLAR KNO_3 54%, $NaNO_3$ 46%)

Temp. t °C	Density ρ kg/m³	Coefficient of Thermal Expansion β 1/K	Kinematic Viscosity ν m²/s	Thermal Diffusivity α m²/s	Prandtl Number Pr	Specific Heat c J/kgK	Thermal Conductivity k W/mK	Absolute Viscosity μ Ns/m²
357	1864	3.40×10^{-3}	1.189×10^{-6}	$.167 \times 10^{-6}$	7.12	1645	0.512	2.217×10^{-3}
367	1858	3.47×10^{-3}	1.130×10^{-6}	$.169 \times 10^{-6}$	6.68	1633	0.513	2.100×10^{-3}
377	1851	3.53×10^{-3}	1.079×10^{-6}	$.172 \times 10^{-6}$	6.29	1621	0.515	1.997×10^{-3}
387	1845	3.60×10^{-3}	1.033×10^{-6}	$.174 \times 10^{-6}$	5.94	1610	0.517	1.906×10^{-3}
397	1838	3.67×10^{-3}	0.994×10^{-6}	$.177 \times 10^{-6}$	5.62	1598	0.519	1.826×10^{-3}
407	1831	3.74×10^{-3}	0.959×10^{-6}	$.179 \times 10^{-6}$	5.34	1586	0.521	1.755×10^{-3}
417	1824	3.81×10^{-3}	0.928×10^{-6}	$.182 \times 10^{-6}$	5.09	1574	0.523	1.692×10^{-3}
427	1817	3.88×10^{-3}	0.900×10^{-6}	$.185 \times 10^{-6}$	4.87	1562	0.525	1.636×10^{-3}
437	1810	3.95×10^{-3}	0.876×10^{-6}	$.188 \times 10^{-6}$	4.66	1551	0.527	1.585×10^{-3}
447	1803	4.02×10^{-3}	0.854×10^{-6}	$.191 \times 10^{-6}$	4.48	1539	0.529	1.540×10^{-3}

1 W/mK = 0.86 kcal/m hr °C   1 J/kg K = 238.9 × 10⁻⁶ kcal/kg °C   1 Ns/m² = 0.102 kgfs/m²

PROPERTY VALUES OF ORGANIC COMPOUNDS AT 20°C

Liquid	Chemical Formula	Density ρ kg/m^3	Coefficient of Thermal Expansion β $1/K$	Kinematic Viscosity ν m^2/s	Thermal Diffusivity α m^2/s	Prandtl Number Pr	Specific Heat c J/kgK	Thermal Conductivity k W/mK	Absolute Viscosity μ Ns/m^2
Acetic acid	$C_2H_4O_2$	1049	1.07×10^{-3}		90.6×10^{-9}		2031	0.193	
Acetone	C_3H_6O	791	1.43×10^{-3}	0.418×10^{-6}	105.4×10^{-9}	3.97	2160	0.180	0.331×10^{-3}
Chloroform	$CHCl_3$	1489	1.28×10^{-3}	0.390×10^{-6}	89.6×10^{-9}	3.97	967	0.129	0.580×10^{-3}
Ethyl acetate	$C_4H_8O_2$	900	1.38×10^{-3}	0.499×10^{-6}	75.7×10^{-9}	3.97	2010	0.137	0.449×10^{-3}
Ethyl alcohol	C_2H_6O	790	1.10×10^{-3}	1.52×10^{-6}	93.3×10^{-9}	3.97	2470	0.182	0.120×10^{-3}
Ethylene glycol	$C_2H_6O_2$	1115		17.8×10^{-6}	97.1×10^{-9}	3.97	2382	0.258	19.900×10^{-3}
Glycerol	$C_3H_8O_3$	1260	0.50×10^{-3}	1175×10^{-6}	93.2×10^{-9}	3.97	2428	0.285	14.80×10^{-3}
n-Heptane	C_7H_{16}	684	1.24×10^{-3}	0.598×10^{-6}	92.2×10^{-9}	3.97	2219	0.140	0.409×10^{-3}
n-Hexane	C_6H_{14}	660	1.35×10^{-3}	0.485×10^{-6}	110.2×10^{-9}	3.97	1884	0.137	0.320×10^{-3}
Isobutyl alcohol	$C_4H_{10}O$	804	0.94×10^{-3}	4.91×10^{-6}	72.4×10^{-9}	3.97	2303	0.134	3.950×10^{-3}
Methyl alcohol	CH_4O	792	1.19×10^{-3}	0.737×10^{-6}	108.4×10^{-9}	3.97	2470	0.212	0.584×10^{-3}
n-Octane	C_8H_{18}	720	1.14×10^{-3}	0.750×10^{-6}	93.8×10^{-9}	3.97	2177	0.147	0.540×10^{-3}
n-Pentane	C_5H_{12}	626	1.60×10^{-3}	0.366×10^{-6}	99.8×10^{-9}	3.97	2177	0.136	0.229×10^{-3}
Toluene	C_7H_8	866	1.08×10^{-3}	0.677×10^{-6}	104.1×10^{-9}	3.97	1675	0.151	0.586×10^{-3}
Turpentine	$C_{10}H_{16}$	855	0.97×10^{-3}	1.74×10^{-6}	83.2×10^{-9}	3.97	1800	0.128	1.487×10^{-3}

$1\ W/mK = 0.86\ kcal/m\ hr\ °C \qquad 1\ J/kg\ K = 238.9 \times 10^{-6}\ kcal/kg\ °C \qquad 1\ Ns/m^2 = 0.102\ kgfs/m^2$

PROPERTY VALUES OF HEAVY WATER (D_2O)

Molecular Weight : 20.033
Freezing Point : 3.82°C
Boiling Point at 9.81×10^4 N/m² : 101.43°C
Critical Temperature : 371.5°C
Critical Pressure : 21.44×10^6 N/m²

Temperature, °C	Enthalpy, h kJ/kg	Specific Heat c, J/kgK	Viscosity of Heavy Water, D_2O / Viscosity of Light Water, H_2O $\dfrac{\mu D_2O}{\mu H_2O}$
3.82	2321.7	—	—
5.0	—	—	1.309
10	2229.9	4226	1.286
15	—	4218	1.267
20	—	4211	1.249
25	2266.7	4207	1.232
30	—	4205	1.215
35	—	4201	—
40	2224.7	4201	—
45	—	4201	—
50	—	4203	—
60	2176.0	—	—
80	2126.3	—	1.164
100	2074.7	—	1.154
120	2020.7	—	—
125	—	—	1.146
140	1950.9	—	—
160	1903.3	—	—
180	1838.7	—	—
200	1769.4	—	—
220	1693.7	—	—

COEFFICIENT OF CUBICAL EXPANSION, β AT 20°C

Fluid	β, 1/K
Water	0.21×10^{-3}
Ammonia	2.46×10^{-3}
Carbon dioxide	14.00×10^{-3}
Glycerine	0.50×10^{-3}
Sulphur dioxide	1.95×10^{-3}
Ethylene glycol	0.65×10^{-3}
Engine oil	0.70×10^{-3}
Dichloro difluro methane (Refrigerant-12) CCl_2F_2	3.10×10^{-3}

For values at other temperatures, $\beta = (\Delta\rho)/(\rho \times \Delta T)$ (use $\Delta\rho$, ρ, ΔT from property tables)

Temperature (°C)	Air β, 1/K	Water β, 1/K
5	3.60×10^{-3}	0.07×10^{-3}
10	3.54×10^{-3}	0.11×10^{-3}
20	3.42×10^{-3}	0.21×10^{-3}
30	3.30×10^{-3}	0.31×10^{-3}
40	3.19×10^{-3}	0.41×10^{-3}
50	3.10×10^{-3}	0.48×10^{-3}
60	3.00×10^{-3}	0.53×10^{-3}
70	2.91×10^{-3}	0.58×10^{-3}
80	2.83×10^{-3}	0.64×10^{-3}
90	2.75×10^{-3}	0.70×10^{-3}
100	2.68×10^{-3}	0.76×10^{-3}
150	2.36×10^{-3}	1.01×10^{-3}
200	2.11×10^{-3}	1.32×10^{-3}
250	1.91×10^{-3}	2.92×10^{-3}
300	1.75×10^{-3}	3.54×10^{-3}
500	1.29×10^{-3}	—
1000	0.77×10^{-3}	—
2000	0.44×10^{-3}	—

For gases, $\beta = \dfrac{1}{T}$, T in K.

HEAT AND MASS TRANSFER DATA BOOK

VARIATION OF THERMAL CONDUCTIVITY OF LIQUIDS WITH TEMPERATURE

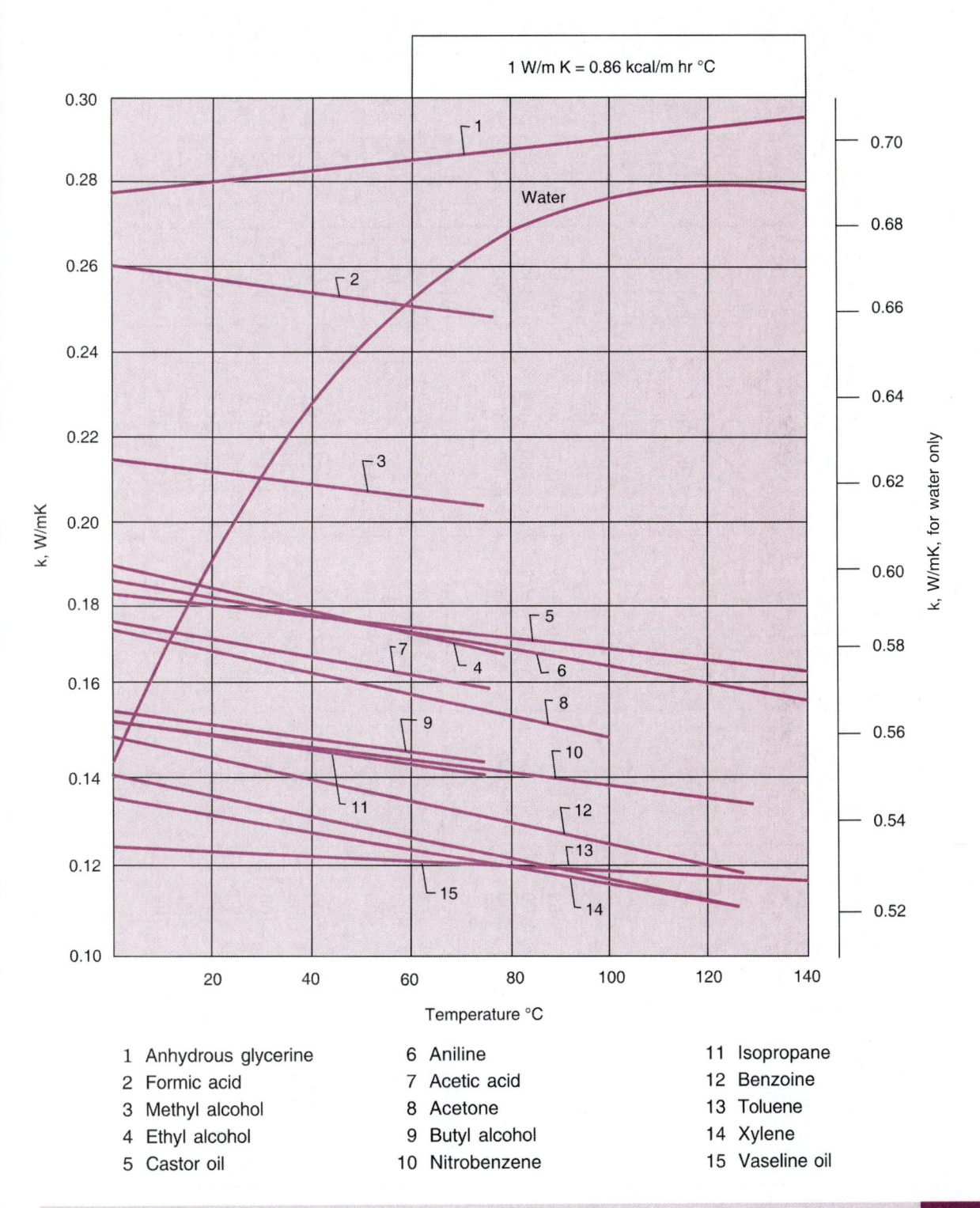

1 W/m K = 0.86 kcal/m hr °C

k, W/mK

k, W/mK, for water only

Temperature °C

1 Anhydrous glycerine
2 Formic acid
3 Methyl alcohol
4 Ethyl alcohol
5 Castor oil
6 Aniline
7 Acetic acid
8 Acetone
9 Butyl alcohol
10 Nitrobenzene
11 Isopropane
12 Benzoine
13 Toluene
14 Xylene
15 Vaseline oil

PHYSICAL PROPERTIES OF LIQUID METALS

Name of Metal t_m Melting Point t_b Boiling Point	Temp. t °C	Density ρ kg/m^3	Kinematic Viscosity ν m^2/s	Thermal Diffusivity α m^2/s	Prandtl Number Pr	Specific Heat c J/kgK	Thermal Conductivity k W/mK
MERCURY, Hg $t_m = -38.9°C$ $t_b = 357°C$	20	13550	0.114×10^{-6}	4.361×10^{-6}	0.0272	139.0	7.91
	100	13350	0.094×10^{-6}	4.889×10^{-6}	0.0192	137.3	8.96
	150	13230	0.086×10^{-6}	5.306×10^{-6}	0.0162	137.3	9.65
	200	13120	0.080×10^{-6}	5.722×10^{-6}	0.0140	137.3	10.35
	300	12880	0.071×10^{-6}	6.639×10^{-6}	0.0107	137.3	11.75
TIN, Sn $t_m = 231.9°C$ $t_b = 2270°C$	250	6980	0.270×10^{-6}	19.167×10^{-6}	0.0141	255.4	34.08
	300	6940	0.240×10^{-6}	19.028×10^{-6}	0.0126	255.4	33.73
	400	6865	0.200×10^{-6}	18.889×10^{-6}	0.0106	255.4	33.15
	500	6790	0.173×10^{-6}	18.750×10^{-6}	0.0092	255.4	32.56
BISMUTH, Bi $t_m = 271°C$ $t_b = 1490°C$	300	10030	0.171×10^{-6}	8.611×10^{-6}	0.0198	150.7	13.03
	400	9910	0.142×10^{-6}	9.722×10^{-6}	0.0146	150.7	14.42
	500	9785	0.122×10^{-6}	10.833×10^{-6}	0.0113	150.7	15.82
	600	9660	0.108×10^{-6}	11.944×10^{-6}	0.0091	150.7	17.21
LITHIUM, Li $t_m = 186°C$ $t_b = 1317°C$	200	515	1.110×10^{-6}	172.22×10^{-6}	0.0643	418.7	37.22
	300	505	0.927×10^{-6}	183.33×10^{-6}	0.0503	418.7	38.96
	400	495	0.817×10^{-6}	202.78×10^{-6}	0.0404	418.7	41.87
	500	484	0.734×10^{-6}	225.00×10^{-6}	0.0328	418.7	45.36
POTASSIUM $t_m = 64°C$ $t_b = 760°C$	149	807.3	0.4608×10^{-6}	69.9×10^{-6}	0.0066	800	45.00
	427	741.7	0.2397×10^{-6}	70.7×10^{-6}	0.0034	750	39.50
	704	674.4	0.1905×10^{-6}	65.5×10^{-6}	0.0029	750	33.10

$1 W/mK = 0.86\ kcal/m\ hr\ °C\quad 1\ J/kg\ K = 238.9 \times 10^{-6}\ kcal/kg\ °C$

PHYSICAL PROPERTIES OF LIQUID METALS (Contd.)

Name of Metal t_m Melting Point t_b Boiling Point	Temperature t °C	Density ρ kg/m³	Kinematic Viscosity ν m²/s	Thermal Diffusivity α m²/s	Prandtl Number Pr	Specific Heat c J/kgK	Thermal Conductivity k W/mK
SODIUM, Na $t_m = 97.3°C$ $t_b = 878°C$	150	916	0.594×10^{-6}	68.333×10^{-6}	0.0087	1357	84.90
	200	903	0.506×10^{-6}	67.778×10^{-6}	0.0075	1327	81.41
	300	878	0.394×10^{-6}	63.056×10^{-6}	0.0063	1281	70.94
	400	854	0.330×10^{-6}	58.889×10^{-6}	0.0056	1273	63.97
	500	829	0.289×10^{-6}	54.167×10^{-6}	0.0053	1273	56.99
ALLOY (Bi-Pb) 56.5% Bi + 43.5% Pb $t_m = 123.5°C$ $t_b = 1600°C$	150	10550	0.289×10^{-6}	6.389×10^{-6}	0.0450	147	9.77
	200	10490	0.243×10^{-6}	6.669×10^{-6}	0.0364	147	10.35
	300	10360	0.187×10^{-6}	7.500×10^{-6}	0.0250	147	11.40
	400	10240	0.157×10^{-6}	10.000×10^{-6}	0.0187	147	12.56
	500	10120	0.136×10^{-6}	9.444×10^{-6}	0.0144	147	13.96
ALLOY (Na-K) 25% Na + 75% K $t_m = -11°C$ $t_b = 184°C$	100	847	0.607×10^{-6}	24.444×10^{-6}	0.0248	1143	23.84
	200	822	0.452×10^{-6}	26.389×10^{-6}	0.0171	1072	23.26
	300	799	0.366×10^{-6}	27.500×10^{-6}	0.0134	1038	22.68
	400	775	0.308×10^{-6}	26.611×10^{-6}	0.0108	1001	22.10
	500	751	0.267×10^{-6}	29.722×10^{-6}	0.0090	967	21.52
ALLOY (Na-K) 45% Na + 55% K $t_m = 19°C$	93	887.4	0.652×10^{-6}	25.556×10^{-6}	0.0260	1130	25.6
	371	821.7	0.287×10^{-6}	31.667×10^{-6}	0.0091	1055	27.5
	704	740.1	0.217×10^{-6}	37.500×10^{-6}	0.0058	1043	28.9

$\beta = (\Delta \rho)/(\rho \times \Delta T)$ (use $\Delta \rho, \rho, \Delta T$ from tables), $\mu = \rho \nu$, $1\,W/mK = 0.86\,kcal/m\,hr\,°C$, $1\,J/kg\,K = 238.9 \times 10^{-6}\,kcal/kg\,°C$

PROPERTY VALUES OF GASES AT ONE ATMOSPHERIC PRESSURE
(use $\beta = 1/T$, T in K)

Fog gases k, c_p, μ and Pr may be taken as not sensitive to pressure. But α, ν, ρ should be corrected for pressure, by calculating the value of ρ at the pressure

Temperature t °C	Density ρ kg/m³	Absolute Viscosity μ Ns/m²	Kinematic Viscosity ν m²/s	Thermal Diffusivity α m²/s	Prandtl Number Pr	Specific Heat c_p J/kgK	Thermal Conductivity k W/mK
DRY AIR							
−50	1.584	14.61 × 10⁻⁶	9.23 × 10⁻⁶	12.644 × 10⁻⁶	0.728	1013	0.02035
−40	1.515	15.20 × 10⁻⁶	10.04 × 10⁻⁶	13.778 × 10⁻⁶	0.728	1013	0.02117
−30	1.453	15.69 × 10⁻⁶	10.80 × 10⁻⁶	14.917 × 10⁻⁶	0.723	1013	0.02198
−20	1.395	16.18 × 10⁻⁶	11.61 × 10⁻⁶	16.194 × 10⁻⁶	0.716	1009	0.02279
−10	1.342	16.67 × 10⁻⁶	12.43 × 10⁻⁶	17.444 × 10⁻⁶	0.712	1009	0.02361
0	1.293	17.16 × 10⁻⁶	13.28 × 10⁻⁶	18.806 × 10⁻⁶	0.707	1005	0.02442
10	1.247	17.65 × 10⁻⁶	14.16 × 10⁻⁶	20.006 × 10⁻⁶	0.705	1005	0.02512
20	1.205	18.14 × 10⁻⁶	15.06 × 10⁻⁶	21.417 × 10⁻⁶	0.703	1005	0.02593
30	1.165	18.63 × 10⁻⁶	16.00 × 10⁻⁶	22.861 × 10⁻⁶	0.701	1005	0.02675
40	1.128	19.12 × 10⁻⁶	16.96 × 10⁻⁶	24.306 × 10⁻⁶	0.699	1005	0.02756
50	1.093	19.61 × 10⁻⁶	17.95 × 10⁻⁶	25.722 × 10⁻⁶	0.698	1005	0.02826
60	1.060	20.10 × 10⁻⁶	18.97 × 10⁻⁶	27.194 × 10⁻⁶	0.696	1005	0.02896
70	1.029	20.59 × 10⁻⁶	20.02 × 10⁻⁶	28.556 × 10⁻⁶	0.694	1009	0.02966
80	1.000	21.08 × 10⁻⁶	21.09 × 10⁻⁶	30.194 × 10⁻⁶	0.692	1009	0.03047
90	0.972	21.48 × 10⁻⁶	22.10 × 10⁻⁶	31.889 × 10⁻⁶	0.690	1009	0.03128
100	0.946	21.87 × 10⁻⁶	23.13 × 10⁻⁶	33.639 × 10⁻⁶	0.688	1009	0.03210
120	0.898	22.85 × 10⁻⁶	25.45 × 10⁻⁶	36.833 × 10⁻⁶	0.686	1009	0.03338
140	0.854	23.73 × 10⁻⁶	27.80 × 10⁻⁶	40.333 × 10⁻⁶	0.684	1013	0.03489
160	0.815	24.52 × 10⁻⁶	30.09 × 10⁻⁶	43.889 × 10⁻⁶	0.682	1017	0.03640
180	0.779	25.30 × 10⁻⁶	32.49 × 10⁻⁶	47.500 × 10⁻⁶	0.681	1022	0.03780
200	0.746	25.99 × 10⁻⁶	34.85 × 10⁻⁶	51.361 × 10⁻⁶	0.680	1026	0.03931
250	0.674	27.36 × 10⁻⁶	40.61 × 10⁻⁶	58.500 × 10⁻⁶	0.677	1038	0.04268
300	0.615	29.71 × 10⁻⁶	48.20 × 10⁻⁶	71.556 × 10⁻⁶	0.674	1047	0.04605
350	0.566	31.38 × 10⁻⁶	55.46 × 10⁻⁶	81.861 × 10⁻⁶	0.676	1059	0.04908
400	0.524	33.05 × 10⁻⁶	63.03 × 10⁻⁶	93.111 × 10⁻⁶	0.678	1067	0.05210
500	0.456	36.19 × 10⁻⁶	79.38 × 10⁻⁶	115.306 × 10⁻⁶	0.687	1093	0.05745
600	0.404	39.13 × 10⁻⁶	96.89 × 10⁻⁶	138.611 × 10⁻⁶	0.699	1114	0.06222
700	0.362	41.78 × 10⁻⁶	115.40 × 10⁻⁶	163.389 × 10⁻⁶	0.706	1135	0.06687
800	0.329	44.33 × 10⁻⁶	134.80 × 10⁻⁶	189.444 × 10⁻⁶	0.713	1156	0.07176
900	0.301	46.68 × 10⁻⁶	155.10 × 10⁻⁶	216.222 × 10⁻⁶	0.717	1172	0.07629
1000	0.277	49.03 × 10⁻⁶	178.00 × 10⁻⁶	246.667 × 10⁻⁶	0.719	1185	0.08071
1100	0.257	51.19 × 10⁻⁶	199.30 × 10⁻⁶	276.250 × 10⁻⁶	0.722	1197	0.08502
1200	0.239	53.45 × 10⁻⁶	223.70 × 10⁻⁶	316.500 × 10⁻⁶	0.724	1210	0.09153

1 W/mK = 0.86 kcal/m hr °C, 1 J/kg K = 238.9 × 10⁻⁶ kcal/kg °C, Ns/m² = 0.102 kgfs/m²

PROPERTY VALUES OF GASES AT ONE ATMOSPHERIC PRESSURE *(Contd.)*

Temperature t °C	Density ρ kg/m^3	Absolute Viscosity μ Ns/m^2	Kinematic Viscosity ν m^2/s	Thermal Diffusivity α m^2/s	Prandtl Number Pr	Specific Heat c_p J/kgK	Thermal Conductivity k W/mK
NITROGEN							
0	1.250	16.67 × 10^{-6}	13.30 × 10^{-6}	19.139 × 10^{-6}	0.705	1030	0.02431
100	0.916	20.69 × 10^{-6}	22.50 × 10^{-6}	32.222 × 10^{-6}	0.678	1034	0.03152
200	0.723	24.22 × 10^{-6}	33.60 × 10^{-6}	50.833 × 10^{-6}	0.656	1043	0.03850
300	0.597	27.65 × 10^{-6}	46.40 × 10^{-6}	70.833 × 10^{-6}	0.652	1059	0.04489
400	0.508	30.89 × 10^{-6}	60.90 × 10^{-6}	92.500 × 10^{-6}	0.659	1080	0.05071
500	0.442	33.93 × 10^{-6}	76.90 × 10^{-6}	114.167 × 10^{-6}	0.672	1105	0.05582
600	0.392	36.87 × 10^{-6}	94.30 × 10^{-6}	136.389 × 10^{-6}	0.689	1130	0.06036
700	0.352	39.62 × 10^{-6}	113.00 × 10^{-6}	158.333 × 10^{-6}	0.710	1151	0.06420
800	0.318	42.27 × 10^{-6}	133.00 × 10^{-6}	181.667 × 10^{-6}	0.734	1168	0.06745
900	0.291	45.01 × 10^{-6}	154.00 × 10^{-6}	203.056 × 10^{-6}	0.762	1189	0.07013
1000	0.268	47.46 × 10^{-6}	177.00 × 10^{-6}	222.778 × 10^{-6}	0.795	1202	0.07234
OXYGEN							
0	1.429	19.42 × 10^{-6}	13.6 × 10^{-6}	18.889 × 10^{-6}	0.720	913	0.02466
100	1.050	24.12 × 10^{-6}	23.1 × 10^{-6}	33.611 × 10^{-6}	0.686	934	0.03291
200	0.826	28.54 × 10^{-6}	34.6 × 10^{-6}	43.333 × 10^{-6}	0.674	963	0.04071
300	0.682	32.46 × 10^{-6}	47.8 × 10^{-6}	70.556 × 10^{-6}	0.673	997	0.04803
400	0.580	36.28 × 10^{-6}	62.8 × 10^{-6}	92.500 × 10^{-6}	0.675	1022	0.05501
500	0.504	40.01 × 10^{-6}	79.6 × 10^{-6}	116.667 × 10^{-6}	0.682	1047	0.06152
600	0.447	43.54 × 10^{-6}	97.8 × 10^{-6}	141.111 × 10^{-6}	0.689	1068	0.06745
700	0.402	46.97 × 10^{-6}	117.0 × 10^{-6}	166.667 × 10^{-6}	0.700	1084	0.07280
800	0.363	50.21 × 10^{-6}	138.0 × 10^{-6}	194.444 × 10^{-6}	0.710	1101	0.07769
900	0.333	53.45 × 10^{-6}	161.0 × 10^{-6}	221.389 × 10^{-6}	0.725	1114	0.08199
1000	0.306	56.49 × 10^{-6}	184.0 × 10^{-6}	250.000 × 10^{-6}	0.738	1122	0.08583
CARBON MONOXIDE							
0	1.250	16.57 × 10^{-6}	13.3 × 10^{-6}	17.944 × 10^{-6}	0.740	1038	0.02326
100	0.916	20.69 × 10^{-6}	22.6 × 10^{-6}	31.389 × 10^{-6}	0.718	1043	0.03030
200	0.723	24.42 × 10^{-6}	33.9 × 10^{-6}	49.722 × 10^{-6}	0.708	1059	0.03652
300	0.596	27.95 × 10^{-6}	47.0 × 10^{-6}	66.111 × 10^{-6}	0.709	1080	0.04257
400	0.508	31.19 × 10^{-6}	61.8 × 10^{-6}	86.389 × 10^{-6}	0.711	1105	0.04850
500	0.442	34.42 × 10^{-6}	78.0 × 10^{-6}	108.056 × 10^{-6}	0.720	1130	0.05408
600	0.392	37.36 × 10^{-6}	96.0 × 10^{-6}	131.667 × 10^{-6}	0.727	1156	0.05966
700	0.351	40.40 × 10^{-6}	115.0 × 10^{-6}	157.222 × 10^{-6}	0.733	1181	0.06501
800	0.317	43.25 × 10^{-6}	135.0 × 10^{-6}	185.278 × 10^{-6}	0.739	1197	0.07013
900	0.291	45.99 × 10^{-6}	157.0 × 10^{-6}	213.333 × 10^{-6}	0.740	1214	0.07548
1000	0.268	48.74 × 10^{-6}	180.0 × 10^{-6}	244.722 × 10^{-6}	0.744	1231	0.08060

1 W/mK = 0.86 kcal/m hr °C,   1 J/kg K = 238.9 × 10^{-6} kcal/kg °C,   1 Ns/m^2 = 0.102 kgfs/m^2

PROPERTY VALUES OF GASES AT ONE ATMOSPHERIC PRESSURE *(Contd.)*

Temperature t °C	Density ρ kg/m³	Absolute Viscosity μ Ns/m²	Kinematic Viscosity ν m²/s	Thermal Diffusivity α m²/s	Prandtl Number Pr	Specific Heat c_p J/kgK	Thermal Conductivity k W/mK
CARBON DIOXIDE							
0	1.977	14.02×10^{-6}	7.09×10^{-6}	9.111×10^{-6}	0.780	816	0.01465
100	1.477	18.24×10^{-6}	12.60×10^{-6}	17.250×10^{-6}	0.733	913	0.02279
200	1.143	22.36×10^{-6}	19.20×10^{-6}	27.306×10^{-6}	0.715	992	0.03094
300	0.944	26.38×10^{-6}	27.30×10^{-6}	39.167×10^{-6}	0.712	1055	0.03908
400	0.802	30.20×10^{-6}	36.70×10^{-6}	53.056×10^{-6}	0.709	1109	0.04722
500	0.698	33.93×10^{-6}	47.20×10^{-6}	68.333×10^{-6}	0.713	1156	0.05489
600	0.618	37.66×10^{-6}	58.30×10^{-6}	85.556×10^{-6}	0.723	1193	0.06210
700	0.555	41.09×10^{-6}	71.40×10^{-6}	101.667×10^{-6}	0.730	1223	0.06885
800	0.502	44.62×10^{-6}	85.30×10^{-6}	120.000×10^{-6}	0.741	1248	0.07513
900	0.460	48.15×10^{-6}	100.00×10^{-6}	138.611×10^{-6}	0.757	1273	0.08094
1000	0.423	51.48×10^{-6}	116.00×10^{-6}	158.056×10^{-6}	0.770	1290	0.08629
SULPHUR DIOXIDE							
0	2.926	12.06×10^{-6}	4.14×10^{-6}	4.722×10^{-6}	0.874	607	0.00837
100	2.140	16.08×10^{-6}	7.51×10^{-6}	8.722×10^{-6}	0.863	662	0.01233
200	1.690	20.01×10^{-6}	11.80×10^{-6}	12.444×10^{-6}	0.856	712	0.01663
300	1.395	23.83×10^{-6}	17.10×10^{-6}	20.139×10^{-6}	0.848	754	0.02128
400	1.185	27.56×10^{-6}	23.30×10^{-6}	27.778×10^{-6}	0.834	783	0.02582
500	1.033	31.28×10^{-6}	30.40×10^{-6}	36.667×10^{-6}	0.822	808	0.03070
600	0.916	35.01×10^{-6}	38.30×10^{-6}	47.222×10^{-6}	0.806	825	0.03580
700	0.892	38.64×10^{-6}	46.80×10^{-6}	59.722×10^{-6}	0.788	837	0.04105
800	0.743	42.17×10^{-6}	56.50×10^{-6}	73.333×10^{-6}	0.776	850	0.04629
900	0.681	45.70×10^{-6}	66.80×10^{-6}	88.889×10^{-6}	0.755	858	0.05187
1000	0.626	49.23×10^{-6}	78.30×10^{-6}	106.111×10^{-6}	0.740	867	0.05757
FLUE GASES							
0	1.295	15.78×10^{-6}	12.20×10^{-6}	16.889×10^{-6}	0.720	1043	0.02279
100	0.950	20.38×10^{-6}	21.54×10^{-6}	30.833×10^{-6}	0.690	1068	0.03128
200	0.748	24.49×10^{-6}	32.80×10^{-6}	48.889×10^{-6}	0.670	1097	0.04012
300	0.617	28.22×10^{-6}	45.81×10^{-6}	70.000×10^{-6}	0.650	1122	0.04838
400	0.525	31.68×10^{-6}	60.38×10^{-6}	94.167×10^{-6}	0.640	1151	0.05699
500	0.457	34.84×10^{-6}	76.30×10^{-6}	121.111×10^{-6}	0.630	1185	0.06559
600	0.405	37.85×10^{-6}	93.61×10^{-6}	150.833×10^{-6}	0.620	1214	0.07419

$1 \, W/mK = 0.86 \, kcal/m \, hr \, °C$ $1 \, J/kg \, K = 238.9 \times 10^{-6} \, kcal/kg \, °C$ $1 Ns/m^2 = 0.102 \, kgfs/m^2$

PROPERTY VALUES OF GASES AT ONE ATMOSPHERIC PRESSURE (Contd.)

Temperature t °C	Density ρ kg/m³	Absolute Viscosity μ Ns/m²	Kinematic Viscosity ν m²/s	Thermal Diffusivity α m²/s	Prandtl Number Pr	Specific Heat c_p J/kgK	Thermal Conductivity k W/mK
FLUE GASES (Contd.)							
700	0.363	40.68×10^{-6}	112.10×10^{-6}	183.889×10^{-6}	0.610	1239	0.08269
800	0.329	43.37×10^{-6}	131.80×10^{-6}	219.722×10^{-6}	0.600	1264	0.09153
900	0.301	45.90×10^{-6}	152.50×10^{-6}	258.056×10^{-6}	0.590	1290	0.10013
1000	0.275	48.35×10^{-6}	174.30×10^{-6}	303.333×10^{-6}	0.580	1306	0.10897
1100	0.257	50.69×10^{-6}	197.10×10^{-6}	345.556×10^{-6}	0.570	1323	0.11746
1200	0.240	52.98×10^{-6}	221.00×10^{-6}	392.500×10^{-6}	0.560	1340	0.12677
ARGON							
0	1.784	21.08×10^{-6}	11.8×10^{-6}	17.800×10^{-6}	0.663	519	0.01651
100	1.305	26.97×10^{-6}	20.6×10^{-6}	31.111×10^{-6}	0.661	519	0.02117
200	1.030	32.17×10^{-6}	31.2×10^{-6}	47.778×10^{-6}	0.653	519	0.02559
300	0.850	36.87×10^{-6}	43.4×10^{-6}	67.778×10^{-6}	0.640	519	0.02989
400	0.724	41.09×10^{-6}	56.7×10^{-6}	90.556×10^{-6}	0.628	519	0.03396
500	0.627	45.21×10^{-6}	72.0×10^{-6}	116.667×10^{-6}	0.619	519	0.03791
600	0.558	48.54×10^{-6}	87.0×10^{-6}	144.167×10^{-6}	0.604	519	0.03943
HELIUM							
0	0.178	18.73×10^{-6}	105×10^{-6}	153.33×10^{-6}	0.684	5204	0.14304
100	0.130	22.95×10^{-6}	176×10^{-6}	263.33×10^{-6}	0.667	5204	0.17910
200	0.103	26.97×10^{-6}	270×10^{-6}	397.22×10^{-6}	0.660	5204	0.21283
300	0.085	30.79×10^{-6}	362×10^{-6}	552.78×10^{-6}	0.656	5204	0.24423
400	0.072	34.32×10^{-6}	474×10^{-6}	730.56×10^{-6}	0.648	5204	0.27563
500	0.063	37.56×10^{-6}	611×10^{-6}	933.33×10^{-6}	0.642	5204	0.30471
600	0.056	40.31×10^{-6}	723×10^{-6}	1144.44×10^{-6}	0.631	5204	0.33262
HYDROGEN							
0	0.0899	8.36×10^{-6}	93×10^{-6}	135.00×10^{-6}	0.688	14069	0.17212
100	0.0657	10.30×10^{-6}	157×10^{-6}	231.67×10^{-6}	0.677	14482	0.21981
200	0.0519	12.06×10^{-6}	233×10^{-6}	350.00×10^{-6}	0.666	14504	0.26400
300	0.0428	13.83×10^{-6}	323×10^{-6}	494.44×10^{-6}	0.655	14533	0.30703
400	0.0364	15.40×10^{-6}	423×10^{-6}	655.56×10^{-6}	0.644	14581	0.34774
500	0.0317	16.77×10^{-6}	534×10^{-6}	833.33×10^{-6}	0.640	14662	0.38728
600	0.0281	18.34×10^{-6}	656×10^{-6}	1027.78×10^{-6}	0.635	14779	0.42682
700	0.0252	19.71×10^{-6}	785×10^{-6}	1230.56×10^{-6}	0.637	14930	0.46287

1 W/mK = 0.86 kcal/m hr °C *1 J/kg K = 238.9 × 10⁻⁶ kcal/kg °C* *1Ns/m² = 0.102 kgfs/m²*

PROPERTY VALUES OF GASES AT ONE ATMOSPHERIC PRESSURE (Contd.)

Temperature t °C	Density ρ kg/m³	Absolute Viscosity μ Ns/m²	Kinematic Viscosity ν m²/s	Thermal Diffusivity α m²/s	Prandtl Number Pr	Specific Heat c_p J/kgK	Thermal Conductivity k W/mK
HYDROGEN (Contd.)							
800	0.0228	21.08 × 10⁻⁶	924 × 10⁻⁶	1452.78 × 10⁻⁶	0.638	15115	0.50009
900	0.0209	22.36 × 10⁻⁶	1070 × 10⁻⁶	1675.00 × 10⁻⁶	0.640	15312	0.53614
1000	0.0192	23.73 × 10⁻⁶	1230 × 10⁻⁶	1911.11 × 10⁻⁶	0.644	15518	0.57103
AMMONIA							
0	0.771	9.36 × 10⁻⁶	12.2 × 10⁻⁶	13.361 × 10⁻⁶	0.908	2043	0.02105
100	0.564	13.04 × 10⁻⁶	23.2 × 10⁻⁶	27.167 × 10⁻⁶	0.852	2219	0.03396
200	0.445	16.67 × 10⁻⁶	38.0 × 10⁻⁶	45.833 × 10⁻⁶	0.818	2399	0.04885
300	0.368	20.59 × 10⁻⁶	56.4 × 10⁻⁶	68.889 × 10⁻⁶	0.812	2583	0.06548
400	0.313	24.32 × 10⁻⁶	78.7 × 10⁻⁶	97.500 × 10⁻⁶	0.796	2747	0.08397
500	0.272	28.15 × 10⁻⁶	105.0 × 10⁻⁶	130.556 × 10⁻⁶	0.793	2918	0.10362
600	0.241	31.97 × 10⁻⁶	134.0 × 10⁻⁶	168.333 × 10⁻⁶	0.792	3082	0.12444
700	0.221	35.99 × 10⁻⁶	168.0 × 10⁻⁶	210.556 × 10⁻⁶	0.791	3245	0.14770
800	0.196	39.82 × 10⁻⁶	205.0 × 10⁻⁶	257.500 × 10⁻⁶	0.793	3404	0.17096
900	0.179	44.13 × 10⁻⁶	247.0 × 10⁻⁶	308.333 × 10⁻⁶	0.798	3555	0.19655
1000	0.165	47.86 × 10⁻⁶	291.0 × 10⁻⁶	363.889 × 10⁻⁶	0.800	3710	0.22213
TOLUENE—C_7H_8							
0	—	6.61 × 10⁻⁶	—	—	0.748	1030	0.01291
100	—	8.86 × 10⁻⁶	—	—	—	1411	—
200	2.38	11.01 × 10⁻⁶	4.65 × 10⁻⁶	—	—	1750	—
300	1.96	13.24 × 10⁻⁶	6.75 × 10⁻⁶	—	—	2047	—
400	1.667	15.40 × 10⁻⁶	9.23 × 10⁻⁶	—	—	2294	—
500	1.45	17.46 × 10⁻⁶	12.00 × 10⁻⁶	—	—	2504	—
600	1.28	19.61 × 10⁻⁶	15.30 × 10⁻⁶	—	—	2671	—
CARBON TETRACHLORIDE (REFRIGERANT-10) CCl_4							
0	—	9.24 × 10⁻⁶	—	—	0.802	520	0.00599
100	5.020	12.31 × 10⁻⁶	2.45 × 10⁻⁶	2.944 × 10⁻⁶	0.828	588	0.00873
200	3.970	15.30 × 10⁻⁶	3.86 × 10⁻⁶	4.722 × 10⁻⁶	0.816	621	0.01163
300	3.275	18.24 × 10⁻⁶	5.59 × 10⁻⁶	7.000 × 10⁻⁶	0.796	641	0.01465
400	2.790	21.18 × 10⁻⁶	7.64 × 10⁻⁶	9.750 × 10⁻⁶	0.776	654	0.01779
500	2.420	24.03 × 10⁻⁶	9.96 × 10⁻⁶	13.111 × 10⁻⁶	0.758	667	0.02117
600	2.150	26.87 × 10⁻⁶	12.60 × 10⁻⁶	16.889 × 10⁻⁶	0.741	676	0.02454

1 W/mK = 0.86 kcal/m hr °C, 1 J/kg K = 238.9 × 10⁻⁶ kcal/kg °C, 1 Ns/m² = 0.102 kgfs/m²

HEAT AND MASS TRANSFER DATA BOOK

PROPERTY VALUES OF GASES AT ONE ATMOSPHERIC PRESSURE *(Contd.)*

Temperature t °C	Density ρ kg/m³	Absolute Viscosity μ Ns/m²	Kinematic Viscosity ν m²/s	Thermal Diffusivity α m²/s	Prandtl Number Pr	Specific Heat c_p J/kgK	Thermal Conductivity k W/mK
ACETONE—C_3H_6O							
0	—	6.86×10^{-6}	—	—	0.886	1256	0.00972
100	1.870	9.41×10^{-6}	5.07×10^{-6}	6.056×10^{-6}	0.840	1537	0.01733
200	1.470	12.06×10^{-6}	8.22×10^{-6}	10.222×10^{-6}	0.806	1788	0.02687
300	1.220	14.71×10^{-6}	12.10×10^{-6}	15.667×10^{-6}	0.774	2022	0.03861
400	1.030	17.36×10^{-6}	16.90×10^{-6}	22.639×10^{-6}	0.743	2236	0.05210
500	0.901	20.01×10^{-6}	22.30×10^{-6}	30.833×10^{-6}	0.720	2428	0.06745
600	0.799	22.75×10^{-6}	28.30×10^{-6}	40.833×10^{-6}	0.695	2587	0.08467
BENZENE—C_6H_6							
0	—	6.98×10^{-6}	—	—	0.716	943	0.00922
100	2.55	7.21×10^{-6}	3.74×10^{-6}	5.111×10^{-6}	0.554	1335	0.01733
200	2.01	12.09×10^{-6}	5.99×10^{-6}	8.361×10^{-6}	0.719	1676	0.02814
300	1.66	14.64×10^{-6}	8.80×10^{-6}	12.833×10^{-6}	0.668	1957	0.04164
400	1.41	17.20×10^{-6}	12.10×10^{-6}	18.611×10^{-6}	0.652	2183	0.05757
500	1.23	19.76×10^{-6}	15.90×10^{-6}	26.222×10^{-6}	0.614	2369	0.07641
600	1.09	22.31×10^{-6}	20.40×10^{-6}	35.000×10^{-6}	0.585	2524	0.09630
STEAM							
100	0.598	11.96×10^{-6}	20.0×10^{-6}	19.222×10^{-6}	1.08	2135	0.02373
200	0.464	15.89×10^{-6}	30.6×10^{-6}	36.667×10^{-6}	0.94	1976	0.03349
300	0.384	20.01×10^{-6}	44.3×10^{-6}	57.222×10^{-6}	0.91	2014	0.04419
400	0.326	24.32×10^{-6}	60.5×10^{-6}	82.778×10^{-6}	0.90	2073	0.05594
500	0.284	28.64×10^{-6}	78.8×10^{-6}	112.778×10^{-6}	0.90	2135	0.06838
600	0.252	33.15×10^{-6}	99.8×10^{-6}	147.500×10^{-6}	0.89	2206	0.08176
700	0.226	37.85×10^{-6}	122.0×10^{-6}	186.111×10^{-6}	0.90	2273	0.09560
800	0.204	42.56×10^{-6}	147.0×10^{-6}	230.278×10^{-6}	0.91	2345	0.11025
900	0.187	47.46×10^{-6}	174.0×10^{-6}	275.833×10^{-6}	0.92	2416	0.12444
1000	0.172	52.37×10^{-6}	204.0×10^{-6}	330.556×10^{-6}	0.92	2483	0.14072

1 W/mK = 0.86 kcal/m hr °C, 1 J/kg K = 238.9×10^{-6} kcal/kg °C, 1Ns/m² = 0.102 kgfs/m²

PROPERTY VALUES OF SATURATED STEAM (1 bar = 10^5 N/m^2)

Saturation Temperature t °C	Saturation Pressure P bar	Density ρ kg/m³	Absolute Viscosity μ Ns/m² or kg/ms	Kinematic Viscosity ν m²/s	Thermal Diffusivity α m²/s	Prandtl Number Pr	Specific Heat c_p J/kgK	Thermal Conductivity k W/mK
100	1.013	0.598	11.96 × 10⁻⁶	20.000 × 10⁻⁶	18.583 × 10⁻⁶	1.08	2135	0.02373
110	1.433	0.826	12.45 × 10⁻⁶	15.072 × 10⁻⁶	13.833 × 10⁻⁶	1.09	2177	0.02489
120	1.985	1.12	12.85 × 10⁻⁶	11.473 × 10⁻⁶	10.500 × 10⁻⁶	1.09	2206	0.02593
130	2.701	1.50	13.24 × 10⁻⁶	8.827 × 10⁻⁶	7.972 × 10⁻⁶	1.11	2257	0.02687
140	3.614	1.97	13.53 × 10⁻⁶	6.868 × 10⁻⁶	6.131 × 10⁻⁶	1.12	2315	0.02803
150	4.670	2.55	13.93 × 10⁻⁶	5.463 × 10⁻⁶	4.728 × 10⁻⁶	1.15	2395	0.02884
160	6.181	3.26	14.32 × 10⁻⁶	4.393 × 10⁻⁶	3.722 × 10⁻⁶	1.18	2479	0.03012
170	7.920	4.12	14.71 × 10⁻⁶	3.570 × 10⁻⁶	2.939 × 10⁻⁶	1.21	2583	0.03128
180	10.027	5.16	15.10 × 10⁻⁶	2.926 × 10⁻⁶	2.339 × 10⁻⁶	1.25	2709	0.03268
190	12.551	6.40	15.59 × 10⁻⁶	2.436 × 10⁻⁶	1.872 × 10⁻⁶	1.30	2885	0.03419
200	15.549	7.86	15.98 × 10⁻⁶	2.033 × 10⁻⁶	1.492 × 10⁻⁶	1.34	2973	0.03547
210	19.077	9.58	16.37 × 10⁻⁶	1.709 × 10⁻⁶	1.214 × 10⁻⁶	1.41	3098	0.03722
220	23.198	11.60	16.87 × 10⁻⁶	1.454 × 10⁻⁶	0.983 × 10⁻⁶	1.47	3408	0.03896
230	27.976	14.00	17.36 × 10⁻⁶	1.240 × 10⁻⁶	0.806 × 10⁻⁶	1.54	3634	0.04094
240	33.478	16.80	17.50 × 10⁻⁶	1.042 × 10⁻⁶	0.658 × 10⁻⁶	1.61	3881	0.04291
250	39.776	20.00	18.24 × 10⁻⁶	0.912 × 10⁻⁶	0.544 × 10⁻⁶	1.68	4158	0.04512
260	46.943	23.70	18.83 × 10⁻⁶	0.795 × 10⁻⁶	0.453 × 10⁻⁶	1.75	4467	0.04803
270	55.058	28.10	19.32 × 10⁻⁶	0.688 × 10⁻⁶	0.378 × 10⁻⁶	1.82	4815	0.05106
280	64.202	33.20	19.91 × 10⁻⁶	0.600 × 10⁻⁶	0.317 × 10⁻⁶	1.90	5234	0.05489
290	74.461	39.20	20.59 × 10⁻⁶	0.525 × 10⁻⁶	0.261 × 10⁻⁶	2.01	5694	0.05827
300	85.927	46.20	21.28 × 10⁻⁶	0.461 × 10⁻⁶	0.217 × 10⁻⁶	2.13	6280	0.06269
310	98.700	54.60	21.97 × 10⁻⁶	0.402 × 10⁻⁶	0.175 × 10⁻⁶	2.29	7118	0.06838
320	112.890	64.70	22.85 × 10⁻⁶	0.353 × 10⁻⁶	0.142 × 10⁻⁶	2.50	8206	0.07513
330	128.630	77.00	23.93 × 10⁻⁶	0.311 × 10⁻⁶	0.108 × 10⁻⁶	2.86	9881	0.08257
340	146.050	92.80	25.20 × 10⁻⁶	0.272 × 10⁻⁶	0.081 × 10⁻⁶	3.35	12351	0.09304
350	165.350	114.00	26.58 × 10⁻⁶	0.233 × 10⁻⁶	0.058 × 10⁻⁶	4.03	16245	0.10700
360	186.750	144.00	29.13 × 10⁻⁶	0.202 × 10⁻⁶	0.039 × 10⁻⁶	5.23	23027	0.12793
370	210.540	202.00	33.73 × 10⁻⁶	0.167 × 10⁻⁶	0.014 × 10⁻⁶	11.10	56522	0.17096

$1\ W/mK = 0.86\ kcal/m\ hr\ °C \qquad 1\ J/kg\ K = 238.9 \times 10^{-6}\ kcal/kg\ °C \qquad 1 Ns/m^2 = 0.102\ kgfs/m^2$

CONTACT RESISTANCE, 1/h OF TYPICAL SURFACES

Surface type	Roughness μm	Temperature °C	Pressure atm	1/h m² K/W
416 Stainless, ground, air	2.54	90–200	3–25	0.264×10^{-3}
304 Stainless, ground, air	1.14	20	40–70	0.528×10^{-3}
416 Stainless, ground, with 0.254 mm brass shim, air	2.54	30–200	7	0.352×10^{-3}
Aluminium, ground, air	2.54	150	12–25	0.088×10^{-3}
Aluminium, ground, air	0.25	150	12–25	0.018×10^{-3}
Aluminium, ground, with 0.254 mm brass shim, air	2.54	150	12–200	0.123×10^{-3}
Copper, ground, air	1.27	20	12–200	0.007×10^{-3}
Copper, milled, air	3.81	20	10–50	0.018×10^{-3}
Copper, milled, vacuum	0.25	30	7–70	0.088×10^{-3}

THERMAL CONDUCTIVITY OF AIR AT HIGHER PRESSURES, W/mK

Pressure, atm \ Temperature °C	20	100	180
1	0.0258	0.0307	0.0362
100	0.0280	0.0309	0.0367
200	0.0333	0.0372	0.0409
300	0.0385	0.0434	0.0450
400	0.0410	0.0475	0.0485

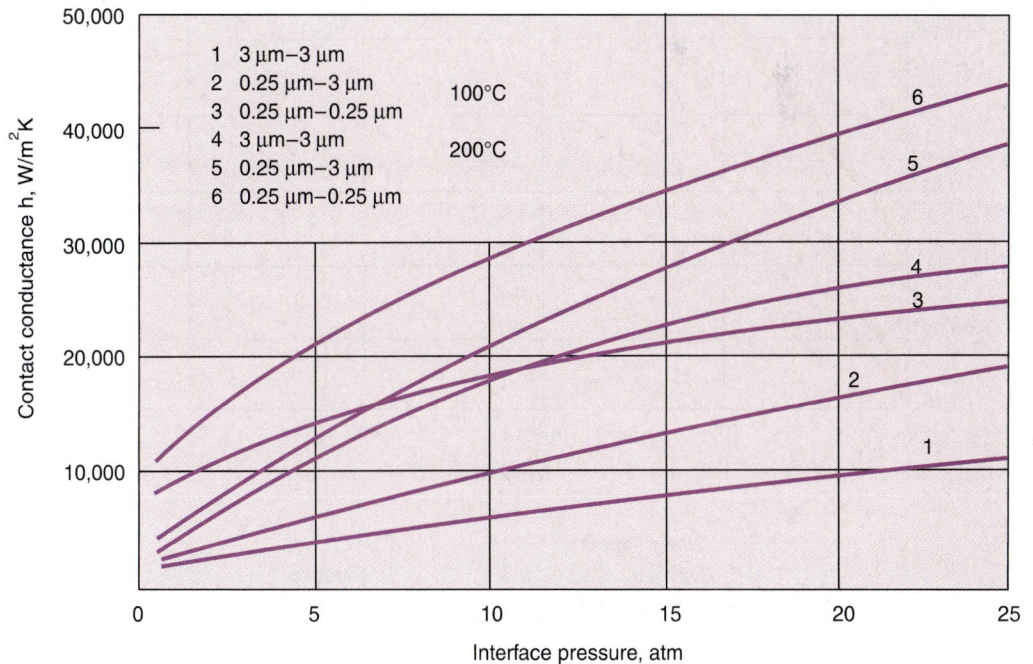

Curves:
1. 3 μm–3 μm
2. 0.25 μm–3 μm
3. 0.25 μm–0.25 μm   100°C
4. 3 μm–3 μm
5. 0.25 μm–3 μm   200°C
6. 0.25 μm–0.25 μm

Contact conductance h, W/m²K vs Interface pressure, atm

Variation of contact conductance of 7075 T6 aluminium joints

VARIATION OF THERMAL CONDUCTIVITY OF GASES WITH TEMPERATURE

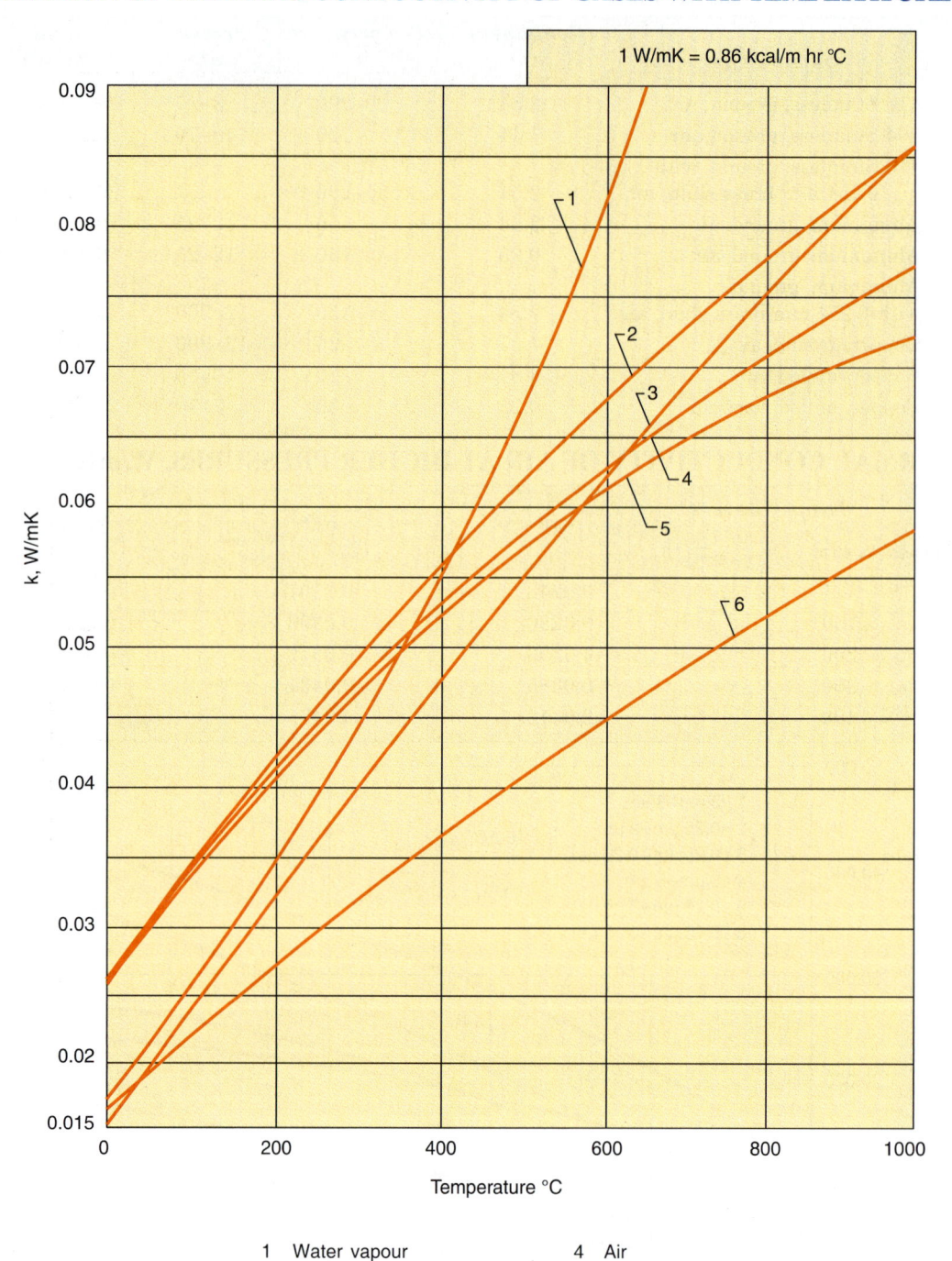

1 W/mK = 0.86 kcal/m hr °C

1. Water vapour
2. Oxygen
3. Carbon dioxide
4. Air
5. Nitrogen
6. Argon

HEAT AND MASS TRANSFER DATA BOOK

VARIATION OF ABSOLUTE VISCOSITY, μ, OF GASES WITH TEMPERATURE

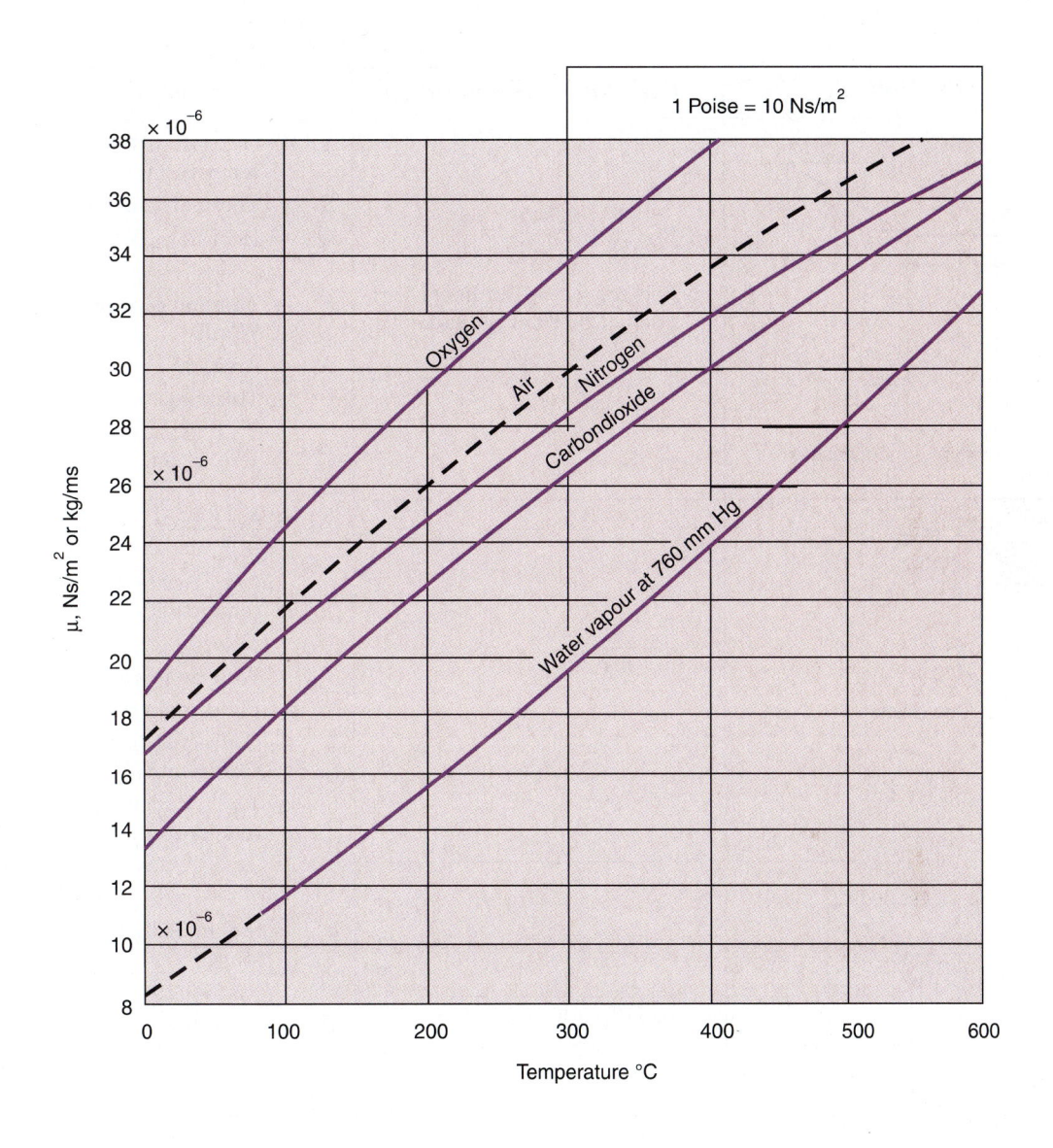

ONE DIMENSIONAL STEADY STATE HEAT TRANSFER

Heat flow, $Q = \dfrac{\Delta T \text{ overall}}{R}$

Description	Correlation and Validity	Notations
Conduction	$R = \dfrac{L}{kA}, \; \dfrac{T_1 - T_x}{T_1 - T_2} = \dfrac{x}{L}$ Replace k by k_m for linear variation of thermal conductivity	ΔT — Overall difference in temperature, °C or K. R — Thermal resistance, K/W Q — Heat flow, W k — Thermal conductivity, W/mK A — Area, m^2 L — Thickness, m T_1, T_x, T_2 temperature at zero, x, and L planes $k_m = k_0 \left(1 \pm \alpha \dfrac{T_1 + T_2}{2}\right)$ α, k_0, constants defining conductivity variation.
Convection	$R = \dfrac{1}{hA}$	T_w — Wall temperature T_∞ — Fluid temperature h — Convective heat transfer coefficient, W/m^{2}K T_1, T_2, Radiating surface temperatures
Radiation-black surfaces	$\dfrac{1}{\sigma A F (T_1^2 + T_2^2)(T_1 + T_2)}$	σ — Stefan Boltzmann Constant $= 5.67 \times 10^{-8}$ W/m^{2}K^4 F = Radiation shape factor, depends on geometry

ONE DIMENSIONAL STEADY STATE HEAT CONDUCTION

Description	Correlation and Validity	Notations
Hollow cylinder	$R = \dfrac{1}{2\pi k L} \ln \dfrac{r_2}{r_1}$ $\dfrac{T_i - T_r}{T_i - T_o} = \dfrac{\ln \dfrac{r}{r_1}}{\ln \dfrac{r_2}{r_1}}$	r_1, r_2, inner and outer radii T_i, T_o, inner and outer surface temperature L, Length of cylinder
Hollow sphere	$R = \dfrac{1}{4\pi k}\left(\dfrac{1}{r_1} - \dfrac{1}{r_2}\right)$ $\dfrac{T_i - T_r}{T_i - T_o} = \dfrac{\dfrac{1}{r_1} - \dfrac{1}{r}}{\dfrac{1}{r_1} - \dfrac{1}{r_2}}$	
Composite plane wall, with convection on sides	$R = \dfrac{1}{A}\left[\dfrac{1}{h_a} + \dfrac{L_1}{k_1} + \dfrac{L_2}{k_2} + \dfrac{L_3}{k_3} + \dfrac{1}{h_b}\right]$	T_a, T_b, fluid temperature outside & inside L_1, L_2, L_3, thickness of layers. For temperature drop in layers use $\Delta T_i = Q \times R_i$ k_1, k_2, k_3, conductivity of layers 1, 2, & 3 h_a, h_b, convection coefficient at a and b, may or may not be present.

Note: ΔK or $\Delta\,°C$ in this chapter and in Convection indicate degree interval and hence identical.

ONE DIMENSIONAL STEADY STATE HEAT CONDUCTION (Contd.)

Description	Correlation and Validity	Notations
Composite cylinder, with convection	$$R = \frac{1}{2\pi L}\left[\frac{1}{h_a r_1} + \frac{1}{k_1}\ln\left(\frac{r_2}{r_1}\right)\right.$$ $$\left. + \frac{1}{k_2}\ln\left(\frac{r_3}{r_2}\right) + \frac{1}{k_3}\ln\left(\frac{r_4}{r_3}\right) + \frac{1}{h_b r_4}\right]$$	r, radius, m L, length, m ΔT, Temperature change in a layer T_a, T_b, fluid temperatures $\Delta T_i = Q \times R_i$
Composite sphere with convection	$$R = \frac{1}{4\pi}\left[\frac{1}{h_i r_i^2} + \frac{1}{k_1}\left[\frac{1}{r_1} - \frac{1}{r_2}\right]\right.$$ $$\left. + \frac{1}{k_2}\left[\frac{1}{r_2} - \frac{1}{r_3}\right] + \frac{1}{h_o r_3^2}\right]$$	T_i, T_o, fluid temperatures $\Delta T_i = Q \times R_i$
Pipe in square section	$$R = \frac{1}{2\pi L}\left[\frac{1}{h_i r} + \frac{L}{k}\ln\frac{1.08a}{2r} + \frac{\pi}{2h_o a}\right]$$	a side of square r radius of cylinder

HEAT AND MASS TRANSFER DATA BOOK

ONE DIMENSIONAL STEADY STATE HEAT CONDUCTION *(Contd.)*

Description	Correlation and Validity	Notations
Pipe with eccentric lagging of length L	$R = \dfrac{1}{2\pi k L} \ln \dfrac{\sqrt{[(r_2+r_1)^2 - e^2]} + \sqrt{[(r_2-r_1)^2 - e^2]}}{\sqrt{[(r_2+r_1)^2 - e^2]} - \sqrt{[(r_2-r_1)^2 - e^2]}}$	T_a, T_b surface temperatures
Series-Parallel Composite Layers Surface area-A	$R = R_1 + R_2 + R_3 + R_4$ $R_1 = R_A$ $R_2 = \dfrac{R_B R_C}{R_B + R_C}$ $R_3 = \dfrac{R_D R_E R_F}{R_D R_E + R_E R_F + R_F R_D}$ $R_4 = R_G$ $A = A_A = A_B + A_C$ $ = A_D + A_E + A_F = A_G$ In case of convection add to R resistances $\dfrac{1}{h_i A}$ and $\dfrac{1}{h_o A}$	$k_A, k_B, \ldots k_G$ —Thermal conductivities of A, B, ... G L—Thicknesses $A_A \ldots A_G$—Area perpendicular to heat flow of materials A, ... G $R_A = \dfrac{L_A}{k_A A_A}$ $R_B = \dfrac{L_B}{k_B A_B}$ $R_C = \dfrac{L_C}{k_C A_C}$ $R_D = \dfrac{L_D}{k_D A_D}$ $R_E = \dfrac{L_E}{k_E A_E}$ $R_F = \dfrac{L_F}{k_F A_F}$ $R_G = \dfrac{L_G}{k_G A_G}$

STEADY STATE CONDUCTION WITH HEAT GENERATION

Description	Correlation and Validity	Notations
Slab	$T_o = T_w + \dfrac{\dot{q}}{2k} L^2$ $\dfrac{T_x - T_o}{T_w - T_o} = (x/L)^2$ $T_x = T_o - \dfrac{\dot{q}}{2k} x^2$ for outside convection $T_w = T_\infty + \dfrac{\dot{q}L}{h}$ $q_x = \dot{q} \times x$	T_∞ Surrounding fluid temp. °C h convection coefficient W/m² K $\dot{q}$ heat generated, W/m³ T_o temp. at mid plane or axis, °C T_x temp. at distance x from mid plane, °C T_∞ Fluid temperature T_w wall temp. °C L half thickness, m
Solid cylinder	$T_o = T_w + \dot{q}R^2/4k$ $\dfrac{T_r - T_w}{T_o - T_w} = 1 - (r/R)^2$ $T_r = T_w + \dfrac{\dot{q}}{4k}(R^2 - r^2)$ $q_r = \pi r^2 L \times \dot{q}$ for outside convection $T_w = T_\infty + \dfrac{\dot{q}R}{2h}$	r, R radii, m T_r temp. at any radius r, °C q_x heat flow at plane x from centre, W/m² q_r heat flow at radius r from centre, W L length of cylinder, m **Critical radius = k/h.**
SLAB Asymmetric boundary condition	$T_x = \dfrac{\dot{q}}{2k}(L^2 - x^2) + \dfrac{x}{2L} \times$ $(T_{w_2} - T_{w_1}) + \dfrac{1}{2}(T_{w_1} + T_{w_2})$ $x_{\max} = \dfrac{k}{2\dot{q}L}(T_{w_2} - T_{w_1})$ $T_{\max} = \dfrac{\dot{q}L^2}{2k} + \dfrac{k}{8\dot{q}L^2} \times$ $(T_{w_1} - T_{w_2})^2 + \dfrac{1}{2}(T_{w_1} + T_{w_2})$	T_{w_1}, T_{w_2} Surface temp. °C L half length $x_{\max}$ Location of maximum Temp. m $T_{\max}$ Maximum Temp. °C

HEAT AND MASS TRANSFER DATA BOOK

STEADY STATE CONDUCTION WITH HEAT GENERATION *(Contd.)*

Description	Correlation and Validity	Notations
Hollow cylinder	Heat flow on both sides $$\frac{T_r - T_i}{T_o - T_i} = \frac{\ln(r/R_i)}{\ln(R_o/R_i)}$$ $$+ \frac{\dot{q}}{4k} \cdot \frac{R_o^2 - R_i^2}{T_o - T_i}$$ $$\times \left[\frac{\ln\frac{r}{R_i}}{\ln r_o/r_i} - \frac{r^2 - R^2}{R_o^2 - R_i^2} \right]$$	See figure R_o, R_i, T_o, T_i, r
Hollow cylinder Inside adiabatic	$$T_r - T_o = \frac{\dot{q}}{4k}[R_o^2 - R_i^2]$$ $$+ \frac{\dot{q}}{2k} R_i^2 \ln \frac{r}{R_o}$$ $$T_i = T_o + \frac{\dot{q}}{4k}(R_o^2 - R_i^2)$$ $$+ \frac{\dot{q}}{2k} R_i^2 \ln \frac{R_i}{R_o}$$ $$q_r = \pi \dot{q} L [r^2 - R_i^2]$$	L Length of cylinder R_i, inside radius R_o, outside radius q_r, heat flow at radius r for L metre, W
Hollow cylinder Outside adiabatic	$$T_r - T_i = \frac{\dot{q}}{2k}(R_i^2 - r^2)$$ $$+ \frac{\dot{q}}{2k} R_o^2 \ln \frac{r}{R_i}$$ $$q_r = \pi q L[R_o^2 - r^2]$$ $$T_o - T_i = \frac{\dot{q}}{2k} \cdot R_o^2 \ln \frac{R_o}{R_i} - \frac{\dot{q}}{4k} \times$$ $$(R_o^2 - R_i^2)$$	
Solid sphere	$$\frac{T_i - T_r}{T_i - T_o} = \left(\frac{r}{R_o}\right)^2,$$ $$T_i = T_o + \frac{\dot{q}}{6k} \cdot R_o^2$$ $$\dot{q}_r = \dot{q}\frac{4}{3}\pi r^3,$$ $$T_r - T_o = \frac{\dot{q}}{6k}(R_o^2 - r^2)$$ for outside convection: $$T_o = T_\infty + \frac{\dot{q}R_o}{3h}, \quad T_r = T_i - \frac{\dot{q}}{6k}r^2$$	q_r heat flow at radius r, W T_i centre temperature, °C T_r temperature at r, °C h convection coefficient on the outside T_o surface temperature, °C T_∞ Outside fluid temperature ***Critical radius*** $= 2k/h$

FINS OR EXTENDED SURFACES

Type of FIN boundary	Temperature distribution $\dfrac{T - T_\infty}{T_b - T_\infty}$	Heat transferred by fin Q	Fin efficiency* η
LONG FIN ($T_L = T_\infty$)	e^{-mx}	$(T_b - T_\infty)(hPkA)^{0.5}$	—
SHORT FIN (end insulated)	$\dfrac{\cosh m(L - X)}{\cosh (mL)}$	$(hPkA)^{0.5}(T_b - T_\infty)\tanh(mL)$	$\dfrac{\tanh(mL)}{mL}$
SHORT FIN (end not insulated)	$\dfrac{\cosh[m(L-X)] + (h_L/mk)\sinh[m(L-X)]}{\cosh(mL) + (h_L/mk)\sinh(mL)}$	$(T_b - T_\infty)\dfrac{\tanh(mL) + (h_L/mk)}{1 + (h_L/mk)\tanh(mL)}(hPkA)^{0.5}$	—
SPECIFIED END TEMPERATURE at $X = L$, $T = T_L$	$\dfrac{\dfrac{T_L - T_\infty}{T_b - T_\infty}\sinh(mX) + \sinh[m(L-X)]}{\sinh(mL)}$	$[(T_b - T_\infty) + (T_L - T_\infty)]\dfrac{\cosh(mL) - 1}{\sinh(mL)}\cdot(hPkA)^{0.5}$	—

* Ratio of the actual heat transferred by fin to the heat transferable by fin, if the entire fin area were at base temperature.

$m = \sqrt{(hP/kA)}$, h_L—convection coefficient at the *tip*

Refer figure for notations of $P, A, L, X, h, T_\infty, T_b$

CIRCUMFERENTIAL, RECTANGULAR, TRIANGULAR FINS

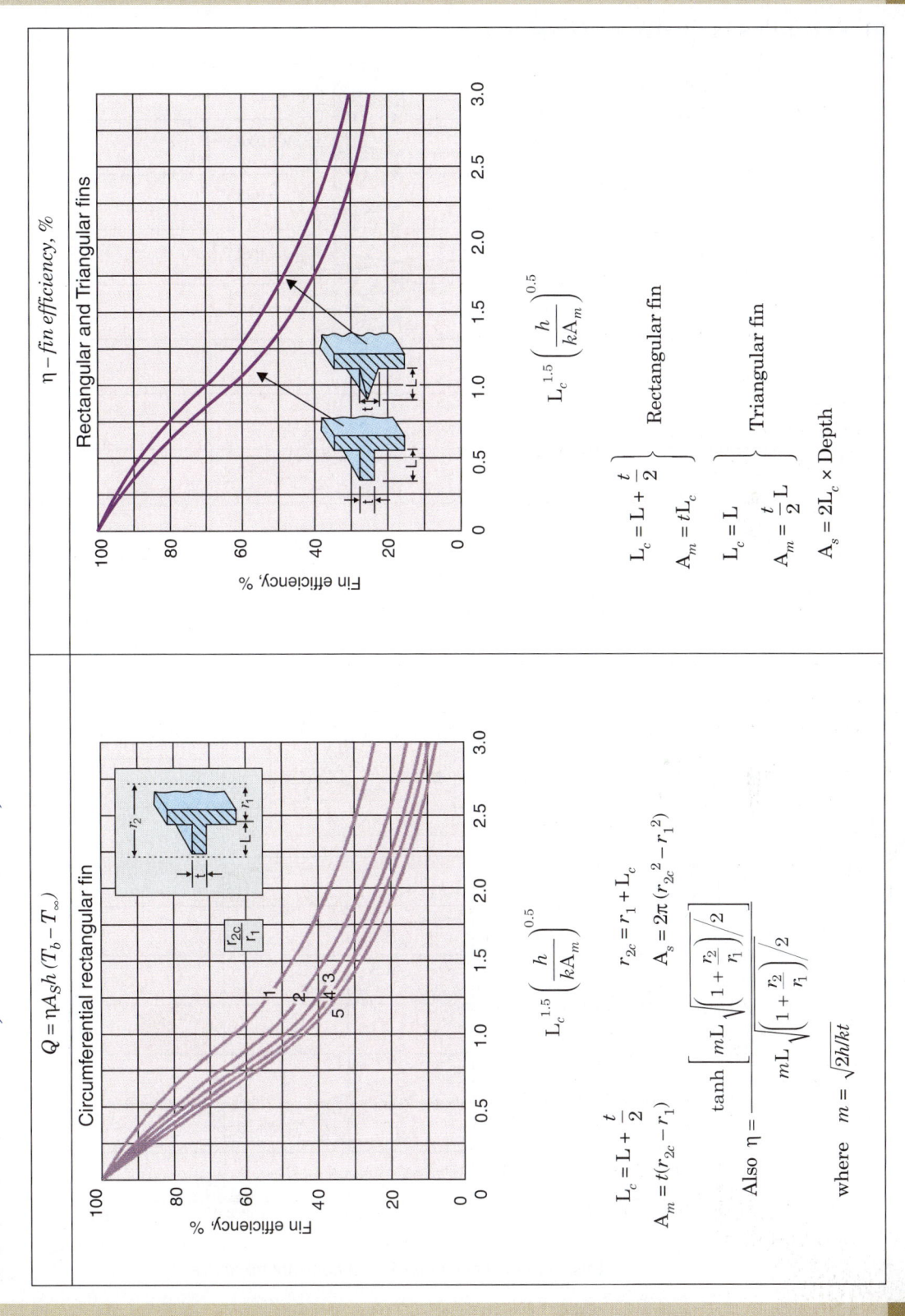

Rectangular and Triangular fins

η – fin efficiency, %

(Chart: Fin efficiency % vs $L_c^{1.5}\left(\dfrac{h}{kA_m}\right)^{0.5}$)

Rectangular fin:
$$L_c = L + \frac{t}{2}$$
$$A_m = tL_c$$

Triangular fin:
$$L_c = L$$
$$A_m = \frac{t}{2}L$$
$$A_s = 2L_c \times \text{Depth}$$

Circumferential rectangular fin

$$Q = \eta A_s h (T_b - T_\infty)$$

(Chart: Fin efficiency % vs $L_c^{1.5}\left(\dfrac{h}{kA_m}\right)^{0.5}$, with curves labeled 1, 2, 3, 4, 5 for parameter $\dfrac{r_{2c}}{r_1}$)

$$L_c = L + \frac{t}{2}$$
$$A_m = t(r_{2c} - r_1)$$
$$r_{2c} = r_1 + L_c$$
$$A_s = 2\pi(r_{2c}^2 - r_1^2)$$

Also
$$\eta = \dfrac{\tanh\left[mL\sqrt{\left(1+\dfrac{r_2}{r_1}\right)/2}\,\right]}{mL\sqrt{\left(1+\dfrac{r_2}{r_1}\right)/2}}$$

where $m = \sqrt{2h/kt}$

PLATE FINS AND DISC FINS

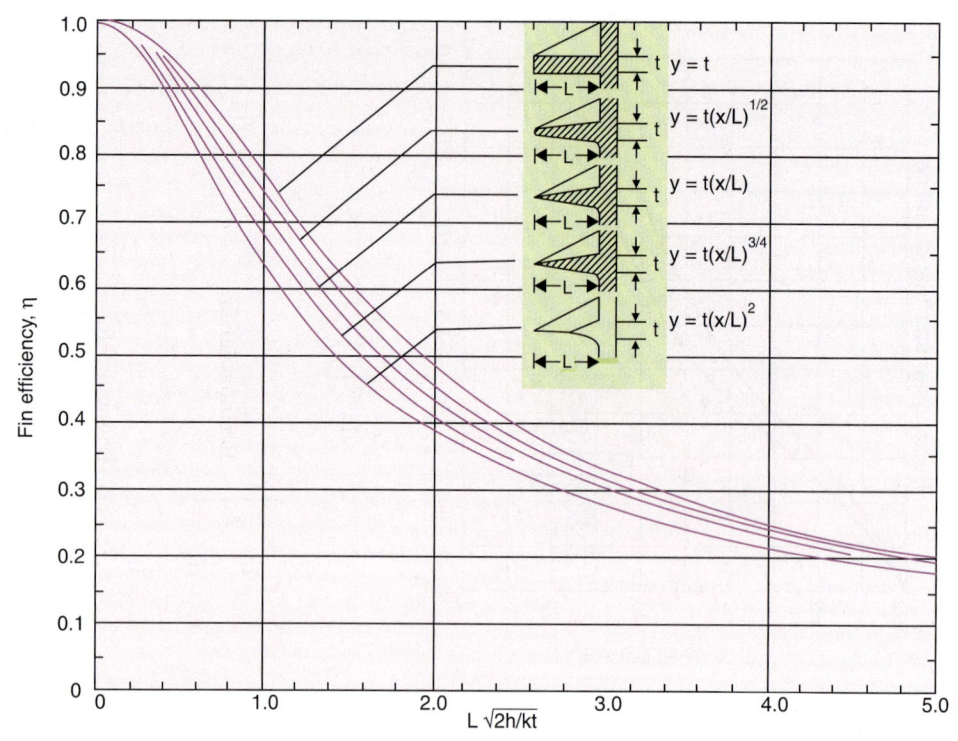

Efficiency of plate fins where the **fin thickness y** varies with the distance x from the root of the fin where $y = t$

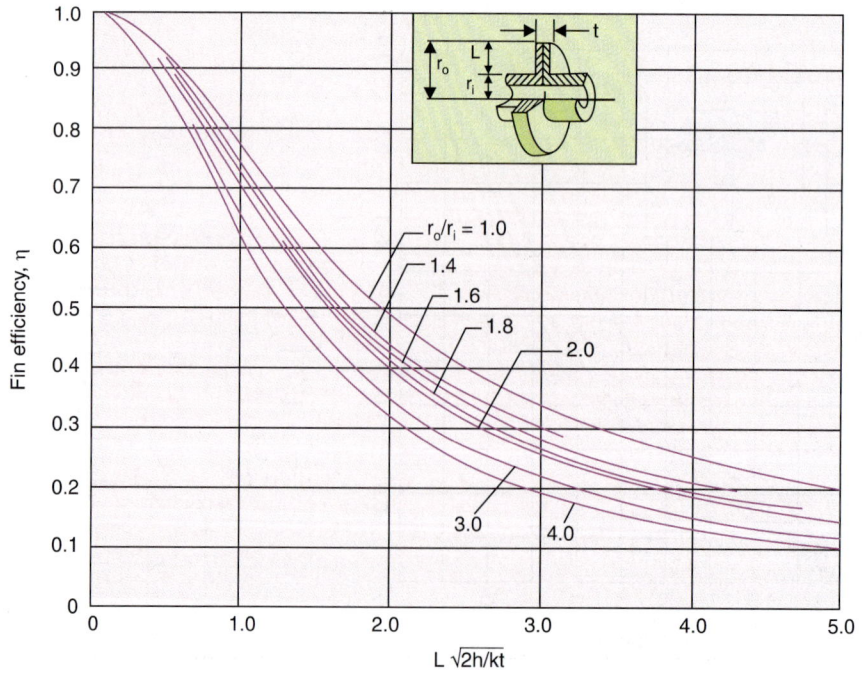

Fin profiles shown:
- $y = t$
- $y = t(x/L)^{1/2}$
- $y = t(x/L)$
- $y = t(x/L)^{3/4}$
- $y = t(x/L)^2$

Axes: Fin efficiency, η vs $L\sqrt{2h/kt}$

Efficiency of circular disk fins of constant thickness

Curves for r_o/r_i = 1.0, 1.4, 1.6, 1.8, 2.0, 3.0, 4.0

Axes: Fin efficiency, η vs $L\sqrt{2h/kt}$

HEAT AND MASS TRANSFER DATA BOOK

SHAPE FACTOR FOR DIFFERENT GEOMETRIES

Formula : Q = $kS\Delta T$	Q — heat transferred, W k — thermal conductivity, W/mK ΔT — overall temperature difference, °C or K S — SHAPE FACTOR, m

SHAPE	SHAPE FACTOR
Buried cylinder of length L in semi infinite medium	Case (*i*) For $r \ll L$ and $D > 3r$: $\dfrac{2\pi L}{\ln \dfrac{2D}{r}}$ Case (*ii*) For $r \ll L$: $\dfrac{2\pi L}{\cosh^{-1}(D/r)}$
Buried sphere in semi infinite medium	$\dfrac{4\pi R}{1 - \left(\dfrac{R}{2D}\right)}$ In infinite medium $S = 4\pi R$
Walls, Edges, Corners (rectangular enclosure)	WALLS : $\dfrac{A}{L}$ EDGES : 0.54 D CORNERS : 0.15 L

SHAPE	SHAPE FACTOR
Heat flow between, buried cylinders of length L	$\dfrac{2\pi L}{\cosh^{-1}\left(\dfrac{D^2 - r_1^2 - r_2^2}{2 r_1^2 r_2^2}\right)}$ $L >> r$ $L >> D$
Vertical cylinder in semi infinite medium (Isothermal)	$\dfrac{2\pi L}{\ln\left(\dfrac{2L}{r}\right)}$ $L >> r$
Rectangular pipe in semi infinite medium (Isothermal)	$1.685 L \left[\log\left(1 + \dfrac{A}{C}\right)\right]^{-0.5} \left(\dfrac{A}{B}\right)^{-0.078}$
Sphere in Semi infinite medium with insulated surface	$\dfrac{4\pi r}{1 + r/2D}$

TWO DIMENSIONAL NODAL EQUATIONS

	Description	Nodal Temperature
	Interior Node	$T_{m,n} = \dfrac{1}{4}[T_{m+1,n} + T_{m,n+1} + T_{m-1,n} + T_{m,n-1}]$
	Convection Boundary	$T_{m,n} = \dfrac{1}{\frac{h\Delta x}{k} + 2}\left[\dfrac{h\Delta x}{k} \cdot T_\infty + T_{m-1,n} \right.$ $\left. + 0.5\, T_{m,n+1} + 0.5\, T_{m,n-1}\right]$
	Exterior Corner with convection boundary	$T_{m,n} = \dfrac{1}{2\left(\dfrac{h\Delta x}{k} + 1\right)} \times$ $\left[\dfrac{2h\Delta x}{k} T_\infty + T_{m-1,n} + T_{m,n-1}\right]$

In case of residual equations use RHS – LHS = 0. It is assumed $\Delta x = \Delta y$

TWO DIMENSIONAL NODAL EQUATIONS *(Contd.)*

	Description	Nodal Temperature
	Interior Corner with convection	$T_{m,n} = \dfrac{1}{2\left(\dfrac{h\Delta x}{k}+3\right)} \left[\dfrac{2h\Delta x}{k} T_\infty + 2T_{m-1,n} + 2T_{m,n+1} + T_{m+1,n} + T_{m,n-1} \right]$
	Insulated Boundary	$T_{m,n} = \dfrac{1}{4} \left[T_{m,n+1} + T_{m,n-1} + 2T_{m-1,n} \right]$
	Curved Boundary	$T_{m,n} = \dfrac{1}{2\left[\dfrac{1}{a}+\dfrac{1}{b}\right]} \left[\dfrac{2}{b(b+1)} T_2 + \dfrac{2}{a+1} T_{m+1,n} + \dfrac{2}{b+1} \cdot T_{m,n-1} + \dfrac{2}{a(a+1)} \cdot T_1 \right]$

In residual equations take $\Delta x = \Delta y$ for unit depth.

TWO DIMENSIONAL NODAL EQUATIONS, TRANSIENT CONDUCTION

m, n, Location, p, p + 1, time interval, $Fo = \dfrac{\alpha \cdot \Delta f}{\Delta x^2}$, $Bi = \dfrac{h\Delta x}{k}$

Node	Nodal Temperature
Internal Node	$T_{m,n}^{p+1} = Fo\,[T_{m-1,n}^p + T_{m,n-1}^p + T_{m+1,n}^p + T_{m,n-1}^p]$ $+ T_{m,n}^p\,[1 - 4Fo]$ $Fo \leq 0.25$
Convection Boundary surface Node	$T_{m,n}^{p+1} = Fo\,[T_{m-1,n}^p + T_{m,n+1}^p + T_{m,n-1}^p + 2.\,Bi\,T_\infty^p]$ $+ [1 - 4Fo - 2\,FoBi]\,T_{m,n}^p$ $Fo\,(2 + Bi) \leq 0.5$
External corner node with convection	$T_{m,n}^{p+1} = 2Fo\,[2T_{m-1,n}^p + T_{m,n+1}^p + 2Bi\,T_\infty^p]$ $+ [1 - 4Fo - 4FoBi]\,T_{m,n}^p$ $Fo\,(1 + Bi) \leq 0.25$
Internal Corner Node with Convection	$T_{m,n}^{p+1} = \dfrac{2}{3}Fo\,[2T_{m,n+1}^p + 2T_{m+1,n}^p + 2T_{m-1,n}^p + T_{m,n-1}^p$ $+ 2Bi\,T_\infty^p] + \left[1 - 9Fo - \dfrac{2}{3}FoBi\right]\cdot T_{m,n}^p$ $Fo\,(3 + Bi) \leq 0.75$
Insulated of surface Node	$T_{m,n}^{p+1} = Fo\,[2T_{m-1,n}^p + T_{m,n+1}^p + T_{m,n-1}^p] + [1 - 4Fo]\,T_{m,n}^p$

TRANSIENT CONDUCTION

Lumped Parameter System

Description	Correlation and Validity	Notations
A body with high thermal conductivity suddenly exposed to a convective atmosphere	Valid for $\dfrac{hL}{k_s} < 0.1$ $$\dfrac{T - T_\infty}{T_o - T_\infty} = \exp\left[-\dfrac{hA_s}{c V \rho}\tau\right]$$ $= \exp[-Bi \cdot Fo]$ Time constant, τ_c = time at which, $\dfrac{T - T_\infty}{T_o - T_\infty} = \dfrac{1}{e}$ $\tau_c = c\rho V / h A_s$ $q = hA_s(T - T_\infty)$ $q_t = \rho c V [T - T_o]$	h — average convective heat transfer coefficient on the solid surface L — Significant length $= \dfrac{V}{A_s}$ V — volume, m^3, A_s — surface area, m^2 k_s — Thermal conductivity of solid τ — time, sec τ_c — time constant, sec T_o — initial temperature of solid, °C T_∞ — surrounding fluid temperature, °C T — temperature of solid at time τ, Bi — Biot number, $\dfrac{hL}{k_s}$ Fo — Fourier Number, $\alpha\tau/L^2$ q — instantaneous heat flow rate, W q_t — total heat flow upto time τ, J c — Specific heat of solid, J/kgK ρ — Density of solid, kg/m^3 α — Thermal Diffusivity of solid, m^2/s
h_1, A_1, T_1 h_2, A_2, T_2 T_∞ ② V_1 ① V_2	$\dfrac{T_1 - T_\infty}{T_o - T_\infty} = \dfrac{m_2}{m_1 - m_2} \cdot \exp(m_1\tau)$ $- \dfrac{m_1}{m_2 - m_1} \exp(m_2 \tau)$ $h_1 A_1 (T_1 - T_2) = -\rho_1 c_1 V_1 \cdot \dfrac{dT_1}{d\tau}$	h_i — contact conductance $m_1 = [-(K_1 + K_2 + K_3) + [(K_1 + K_2 + K_3)^2 - 4K_1K_3]^{0.5}]^{0.5}$ $m_2 = -(K_1 + K_2 + K_3)/2$ $- \dfrac{[(K_1 + K_2 + K_3)^2 - 4K_1K_3]^{0.5}}{2}$ $K_1 = \dfrac{h_1 A_1}{\rho_1 c_1 V_1}$ $K_2 = \dfrac{h_1 A_1}{\rho_2 c_2 V_2}$ $K_3 = \dfrac{h_2 A_2}{\rho_2 c_2 V_2}$

TRANSIENT CONDUCTION
Semi Infinite Solid

Description	Correlation and Validity	Notations
Surface temperature suddenly changed and maintained constant	$\dfrac{T_x - T_o}{T_i - T_o} = \text{erf}\ \dfrac{x}{2\sqrt{\alpha\tau}}$ $q_o = \dfrac{k(T_o - T_i)}{\sqrt{\pi\alpha\tau}}$ $q_x = \dfrac{k(T_o - T_i)}{\sqrt{\pi\alpha\tau}} \exp[-x^2/4\alpha\tau]$ $q_\tau = 2k(T_o - T_i)\sqrt{\dfrac{\tau}{\pi\alpha}}$ $\text{erf}(Z) = \dfrac{2}{\sqrt{\pi}}\left[z - \dfrac{z^3}{3\times 1!} + \dfrac{z^5}{5\times 2!} - \dfrac{z^7}{7\times 3!}\cdots\right]$	Refer charts in p 61, 62 and 63 erf — error function of – refer tables next page T_x — Temperature at a distance x from surface at time τ T_o — Surface temperature T_i — Initial temperature of solid α — Thermal diffusivity of solid, m²/s τ — time q_o — heat flux at surface at time τ, W/m² q_x — heat flux at location x at time τ. W/m² q_τ — Total heat flow into solid upto time τ per unit area, J/m²
Body at T_i Suddenly exposed to constant heat flux q at surface,	q = applied constant heat flux at surface, per unit area $T_{x\tau} - T_i = \dfrac{2q(\alpha\tau/\pi)^{0.5}}{k}\cdot\exp\dfrac{-x^2}{4\alpha\tau}$ $\quad -\dfrac{qx}{k}\left[1 - \text{erf}\ \dfrac{x}{2\sqrt{\alpha\tau}}\right]$ At surface $(T_o/T_i) = 2\left(\dfrac{q}{k}\right)\left(\dfrac{\alpha\tau}{\pi}\right)^{0.5}$	x — location of plane considered from surface T_∞ — fluid temperature h — convective heat transfer coefficient k — thermal conductivity T_o — Surface temperature q — constant heat flux at surface
Body at T_i Surface suddenly exposed to a convective condition with constant fluid temperature T_∞ and convective heat transfer coefficient, h	$\dfrac{T_{x\tau} - T_i}{T_\infty - T_i} = \left[1 - \text{erf}\ \dfrac{x}{2\sqrt{\alpha\tau}}\right]$ $\quad -\exp\left[\dfrac{hx}{k} + \dfrac{h^2\alpha\tau}{k^2}\right]$ $\quad \left[1 - \text{erf}\left(\dfrac{x}{2\sqrt{\alpha\tau}} + \dfrac{h\sqrt{\alpha\tau}}{k}\right)\right]$ Alternately refer charts on page 61, 62 and 63 $q_s = h[T_\infty - T_o']$ At surface $\dfrac{T_o - T_i}{T_\infty - T_i}$ $= 1 - \left[1 - \text{erf}\left(\dfrac{h\sqrt{\alpha\tau}}{k}\right)\right]\cdot\exp\left(\dfrac{h^2\alpha\tau}{k^2}\right)$	q_s — heat flow at surface at time τ, W/m² T_o' — Temperature at $x = 0$, the surface at time τ $T_{x\tau}$ — Temperature at x at time τ

VALUES OF ERROR FUNCTION OF Z

Z	erf(Z)	Z	erf(Z)	Z	erf(Z)	Z	erf(Z)	Z	erf(Z)	Z	erf(Z)	Z	erf(Z)
0.00	0.00000	0.35	0.37938	0.70	0.67780	1.05	0.86244	1.40	0.95228	1.75	0.98667	2.20	0.998137
0.01	0.01128	0.36	0.38933	0.71	0.68467	1.06	0.86614	1.41	0.95385	1.76	0.98719	2.22	0.998308
0.02	0.02256	0.37	0.39921	0.72	0.69143	1.07	0.86977	1.42	0.95538	1.77	0.98769	2.24	0.998464
0.03	0.03384	0.38	0.40901	0.73	0.69810	1.08	0.87333	1.43	0.95686	1.78	0.98817	2.26	0.998607
0.04	0.04511	0.39	0.41874	0.74	0.70468	1.09	0.87680	1.44	0.95830	1.79	0.98864	2.28	0.998738
0.05	0.05637	0.40	0.42839	0.75	0.71116	1.10	0.88020	1.45	0.95970	1.80	0.98909	2.30	0.998857
0.06	0.06762	0.41	0.43797	0.76	0.71754	1.11	0.88353	1.46	0.96105	1.81	0.98952	2.32	0.998966
0.07	0.07886	0.42	0.44747	0.77	0.72382	1.12	0.88679	1.47	0.96237	1.82	0.98994	2.34	0.999065
0.08	0.09008	0.43	0.45689	0.78	0.73001	1.13	0.88997	1.48	0.96365	1.83	0.99035	2.36	0.999155
0.09	0.10128	0.44	0.46622	0.79	0.73610	1.14	0.89308	1.49	0.96490	1.84	0.99074	2.38	0.999237
0.10	0.11246	0.45	0.47548	0.80	0.74210	1.15	0.89612	1.50	0.96610	1.85	0.99111	2.40	0.999311
0.11	0.12362	0.46	0.48466	0.81	0.74800	1.16	0.89910	1.51	0.96728	1.86	0.99147	2.42	0.999379
0.12	0.13476	0.47	0.49374	0.82	0.75381	1.17	0.90200	1.52	0.96841	1.87	0.99182	2.44	0.999441
0.13	0.14587	0.48	0.50275	0.83	0.75952	1.18	0.90484	1.53	0.96952	1.88	0.99216	2.46	0.999497
0.14	0.15695	0.49	0.51167	0.84	0.76514	1.19	0.90761	1.54	0.97059	1.89	0.99248	2.48	0.999547
0.15	0.16800	0.50	0.52050	0.85	0.77067	1.20	0.91031	1.55	0.97162	1.90	0.99279	2.50	0.999593
0.16	0.17901	0.51	0.52924	0.86	0.77610	1.21	0.91296	1.56	0.97263	1.91	0.99309	2.55	0.999689
0.17	0.18999	0.52	0.53790	0.87	0.78144	1.22	0.91553	1.57	0.97360	1.92	0.99338	2.60	0.999764
0.18	0.20094	0.53	0.54646	0.88	0.78669	1.23	0.91805	1.58	0.97455	1.93	0.99366	2.65	0.999822
0.19	0.21184	0.54	0.55494	0.89	0.79184	1.24	0.92050	1.59	0.97546	1.94	0.99392	2.70	0.999866
0.20	0.22270	0.55	0.56332	0.90	0.79691	1.25	0.92290	1.60	0.97635	1.95	0.99418	2.75	0.999899
0.21	0.23352	0.56	0.57162	0.91	0.80188	1.26	0.92524	1.61	0.97721	1.96	0.99443	2.80	0.999925
0.22	0.24430	0.57	0.57982	0.92	0.80677	1.27	0.92751	1.62	0.97804	1.97	0.99466	2.85	0.999944
0.23	0.25502	0.58	0.58792	0.93	0.81156	1.28	0.92973	1.63	0.97884	1.98	0.99489	2.90	0.999959
0.24	0.26570	0.59	0.59594	0.94	0.81627	1.29	0.93190	1.64	0.97962	1.99	0.99511	2.95	0.999970
0.25	0.27633	0.60	0.60386	0.95	0.82089	1.30	0.93401	1.65	0.98038	2.00	0.995322	3.00	0.999978
0.26	0.28690	0.61	0.61168	0.96	0.82542	1.31	0.93606	1.66	0.98110	2.02	0.995720	3.20	0.999994
0.27	0.29742	0.62	0.61941	0.97	0.82987	1.32	0.93806	1.67	0.98181	2.04	0.996086	3.40	0.999998
0.28	0.30788	0.63	0.62705	0.98	0.83423	1.33	0.94002	1.68	0.98249	2.06	0.996424	3.60	1.000000
0.29	0.31828	0.64	0.63459	0.99	0.83851	1.34	0.94191	1.69	0.98315	2.08	0.996734		
0.30	0.32863	0.65	0.64203	1.00	0.84270	1.35	0.94376	1.70	0.98379	2.10	0.997020		
0.31	0.33891	0.66	0.64938	1.01	0.84681	1.36	0.94556	1.71	0.98441	2.12	0.997284		
0.32	0.34913	0.67	0.65663	1.02	0.85084	1.37	0.94731	1.72	0.98500	2.14	0.997525		
0.33	0.35928	0.68	0.66378	1.03	0.85478	1.38	0.94902	1.73	0.98558	2.16	0.997747		
0.34	0.36936	0.69	0.67084	1.04	0.85865	1.39	0.95067	1.74	0.98613	2.18	0.997951		

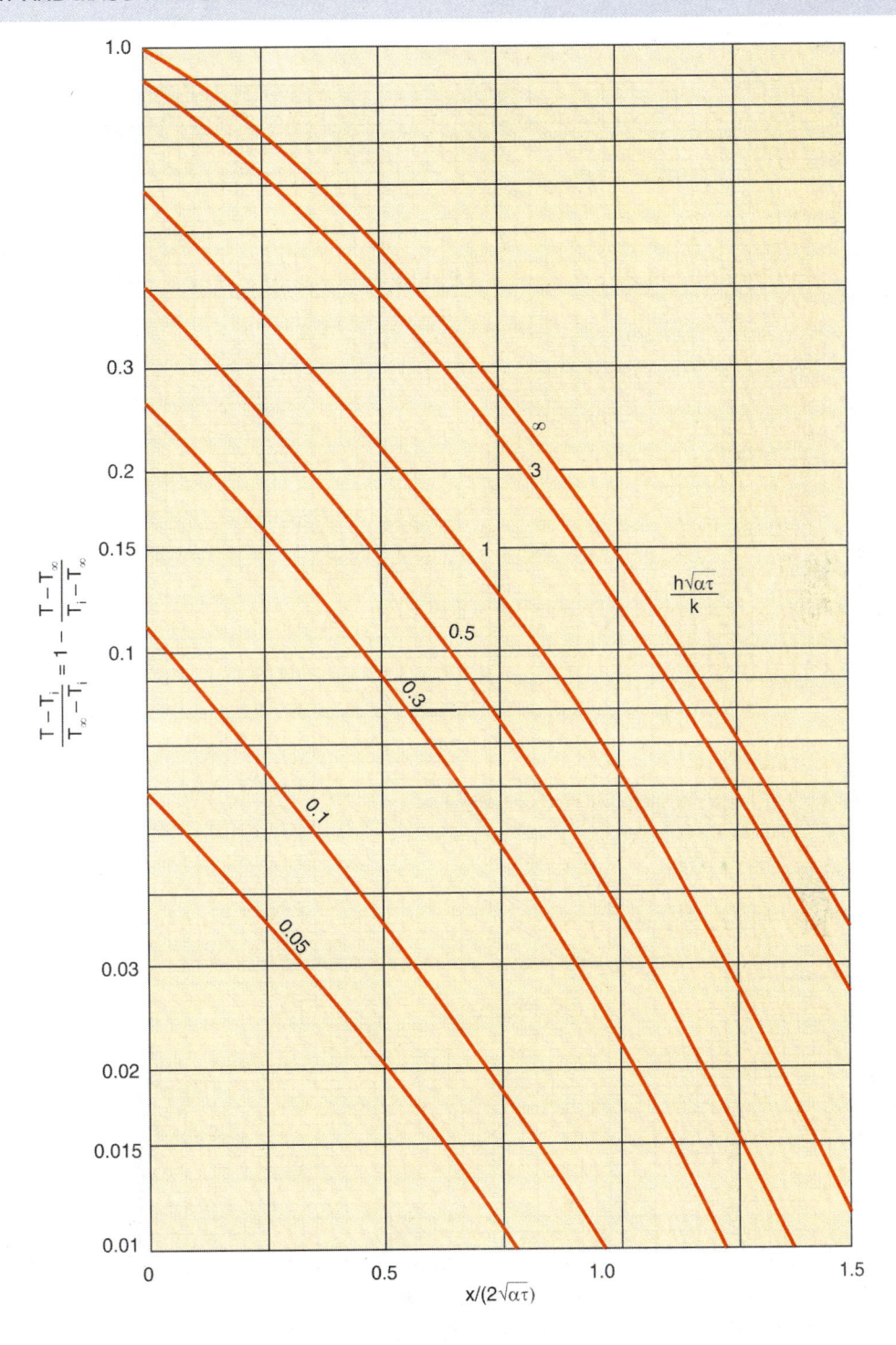

Temperature Distribution in the Semi Infinite Solid with Convection Boundary Condition. $x/(2\sqrt{\alpha\tau}) = 0$ indicates the surface.

SEMI INFINITE SOLID: TEMPERATURE—TIME HISTORY, (LOCATION, x, TEMPERATURE KNOWN, TO FIND TIME)

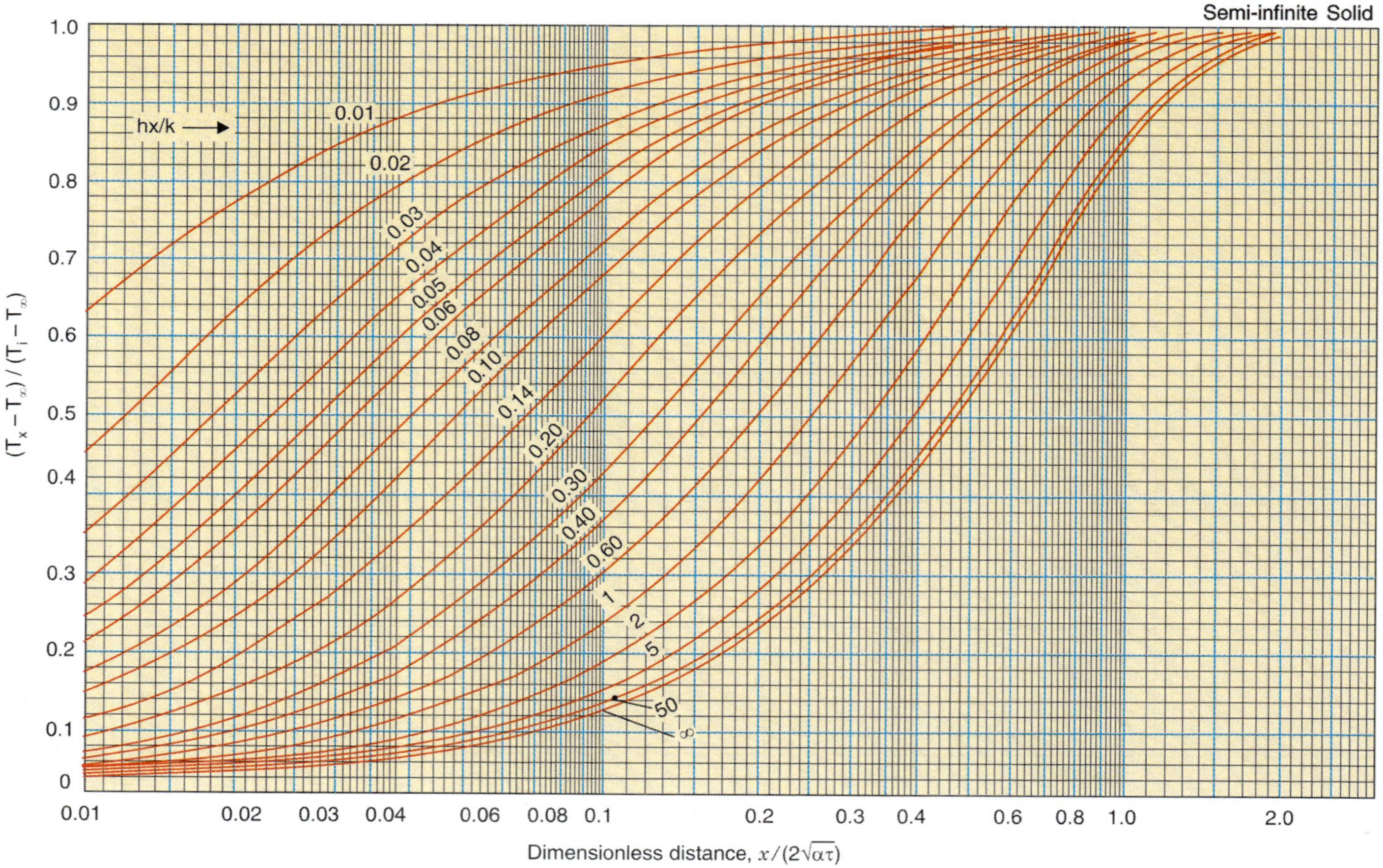

Semi-infinite Solid

HEAT AND MASS TRANSFER DATA BOOK

SEMI INFINITE SOLID: TEMPERATURE–TIME HISTORY, (TEMPERATURE, TIME KNOWN, TO FIND LOCATION, x)

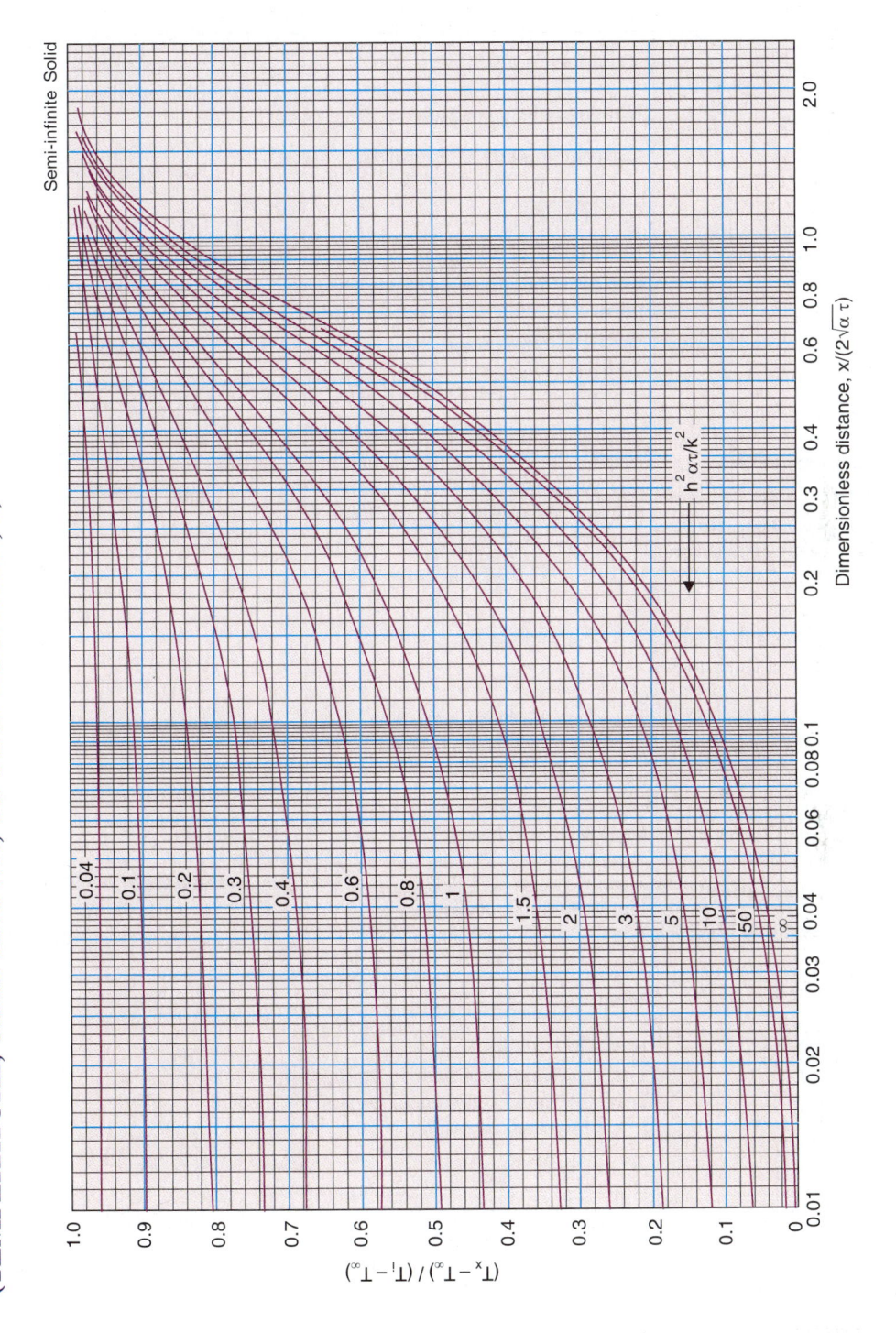

Semi-infinite Solid

Chart: $(T_x - T_\infty)/(T_i - T_\infty)$ versus dimensionless distance $x/(2\sqrt{\alpha\tau})$, with parameter $h^2\alpha\tau/k^2$ taking values 0.04, 0.1, 0.2, 0.3, 0.4, 0.6, 0.8, 1, 1.5, 2, 3, 5, 10, 50, ∞.

TRANSIENT CONDUCTION
Infinite Solids

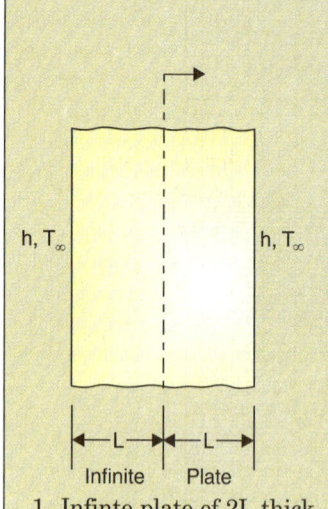

1. Infinite plate of 2L thickness suddenly exposed on both sides to fluid at T_∞, h

Also for one side insulated plane of thickness L.

Heat flow at surface at time τ

Instantaneous heat flow at surface

Valid for Fo > 0.1

Dimensionless position $\dfrac{x}{L}$

For centre plane temperature
$$\dfrac{T_o - T_\infty}{T_i - T_\infty}$$
Refer chart p. 66

For dimensionless position temperature referred to centre line temperature
$$\dfrac{T_{(x/L)} - T_\infty}{T_o - T_\infty} \quad \text{refer chart p. 67}$$

To find x-plane temperature with respect to initial temperature at any time, use the product of the above two expressions i.e.,
$$\dfrac{T_o - T_\infty}{T_i - T_\infty} \cdot \dfrac{T_{(x/L)} - T_\infty}{T_o - T_\infty}$$
$$= \dfrac{T_{(x/L)} - T_\infty}{T_i - T_\infty}$$

For dimensionless heat flow
$$\dfrac{Q}{Q_o} \quad \text{- refer chart p. 68}$$

$q = hA[T_s - T_\infty]$

- $\dfrac{hL}{k}$ — Biot number, Bi
- $\dfrac{\alpha\tau}{L^2}$ — Fourier number, Fo
- T_o — Centre plane temperature at time τ
- T_i — Initial temperature of plate
- T_∞ — Fluid temperature
- $T_{(x/L)}$ — Temperature of plane at x at time τ
- T_s — Surface Temperature at time τ, use $\dfrac{x}{L} = 1$
- Q_o = heat capacity of the body above T_∞ for unit area

  For plate
- Q_o = $\rho cL[T_i - T_\infty]$, J
- Q = Total heat flow upto time τ, J
- q = instantaneous heat flow at surface, W
- T_s = surface temp. at time τ, at $\dfrac{r}{R_o} = 1$

2. Long cylinder, surface Suddenly exposed to fluid at T_∞, h

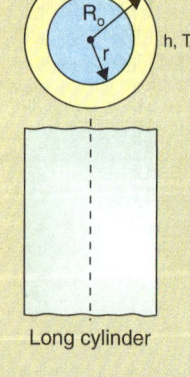

Dimensionless position, $\dfrac{r}{R_o}$

For dimensionless centre line temp. ratio $\dfrac{T_o - T_\infty}{T_i - T_\infty}$, refer chart p. 69

For dimensionless position temperature ratio referred to centre line temperature ratio
$$\dfrac{T_{(r/R_o)} - T_\infty}{T_o - T_\infty}, \text{ refer chart p. 70}$$

For local temperature use the product of the above two to get expression
$$\dfrac{T_{(r/R_o)} - T_\infty}{T_i - T_\infty}$$

- $B_i = \dfrac{hR_o}{k}$
- $F_o = \dfrac{\alpha\tau}{R_o^2}$
- r — any radius
- R_o — outside radius
- $T_{(r/R_o)}$ = Temperature at radius r at time τ

HEAT AND MASS TRANSFER DATA BOOK

TRANSIENT CONDUCTION (Contd.)

Description	Correlation and Validity	Notations
Heat flow upto time τ	$\dfrac{Q}{Q_o}$ – refer chart p. 71	Q — Total heat flow upto time τ per unit length, J
Instantaneous heat flow at surface per unit length	$q = 2\pi h R_o (T_s - T_\infty)$	Q_o — heat capacity above T_∞ per unit length
		$Q_o = \rho c \pi R_o^2 [T_i - T_\infty]$
		T_s — surface temperature at time τ, at $\dfrac{r}{R_o} = 1$
3. Sphere suddenly exposed to fluid at T_∞, h Sphere	Dimensionless position $\dfrac{r}{R_o}$ For dimensionless centre temperature ratio $\dfrac{T_o - T_\infty}{T_i - T_\infty}$ refer chart p. 72 For dimensionless position temperature ratio referred to centre temperature $\dfrac{T_{(r/R_o)} - T_\infty}{T_o - T_\infty}$ refer chart p. 73 For position temperature use the product of the above two expressions $\dfrac{T_{(r/R_o)} - T_\infty}{T_i - T_\infty}$ $= \dfrac{T_o - T_\infty}{T_i - T_\infty} \cdot \dfrac{T_{(r/R_o)} - T_\infty}{T_o - T_\infty}$	$T_{(r/R_o)}$ — Temperature at r at time τ
Total heat flow upto time τ Instantaneous heat flow	For total heat flow $\dfrac{Q}{Q_o}$ – refer chart p. 74 $q = 4\pi R_o^2 h [T_s - T_\infty]$	$Q_o = \dfrac{4}{3} \pi R_o^3 \cdot \rho \cdot c \cdot [T_i - T_\infty]$

INFINITE PLATE: TEMPERATURE—TIME HISTORY AT MID PLANE

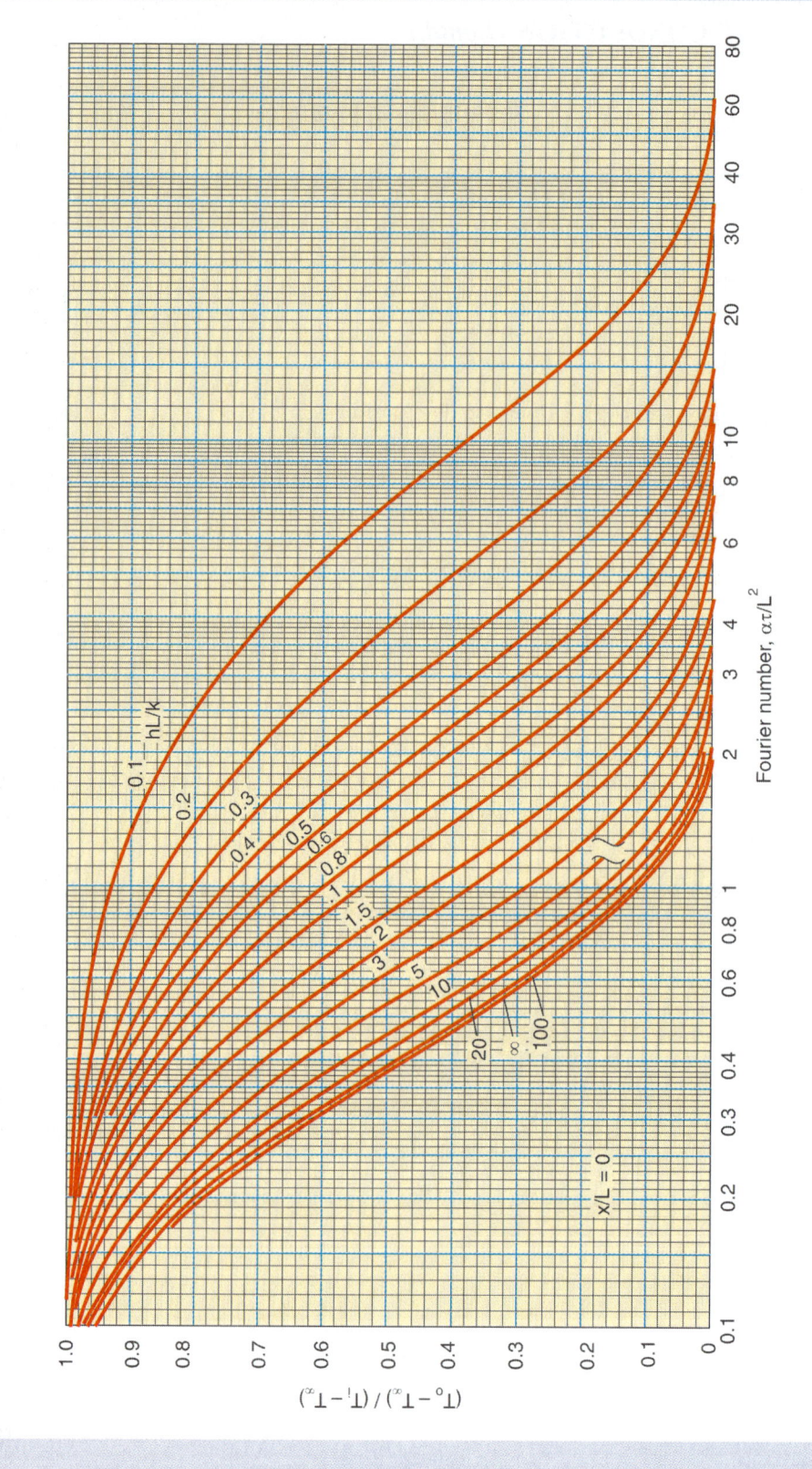

HEAT AND MASS TRANSFER DATA BOOK

INFINITE PLATE : TEMPERATURE—TIME HISTORY AT ANY POSITION, x

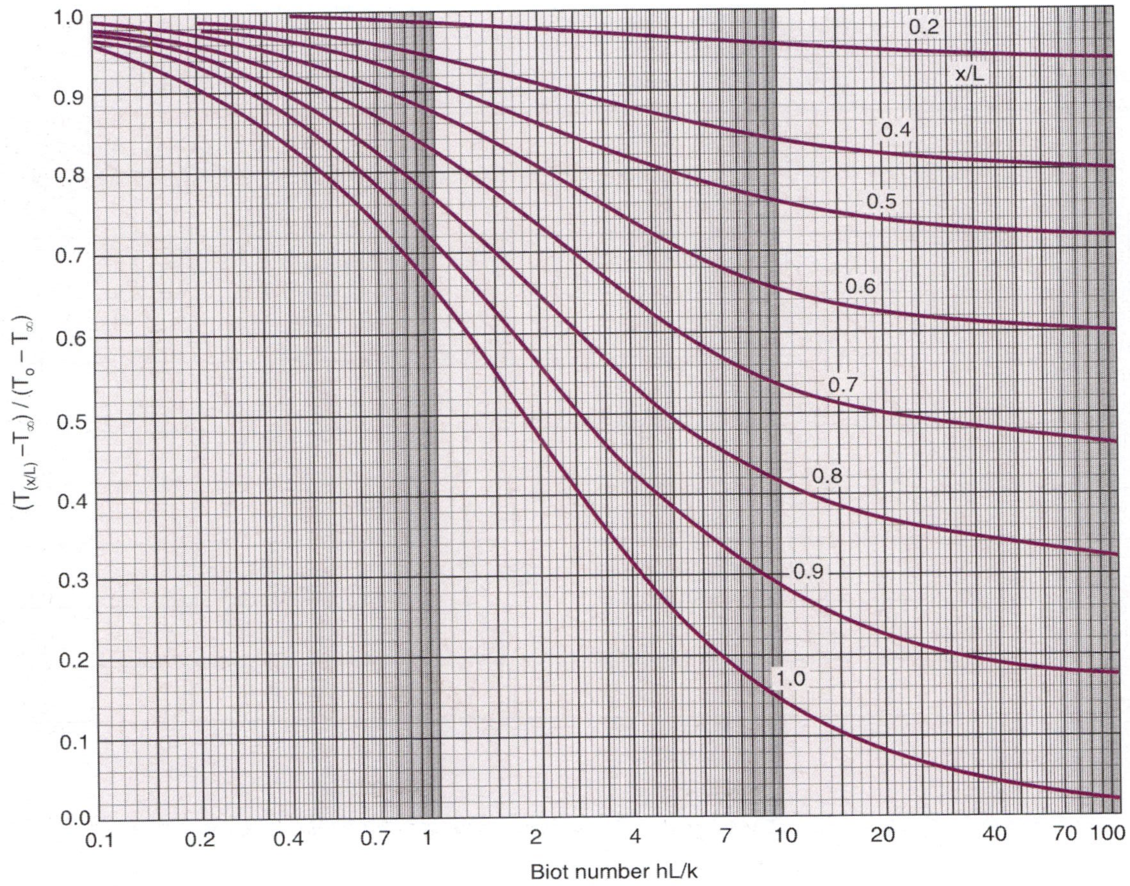

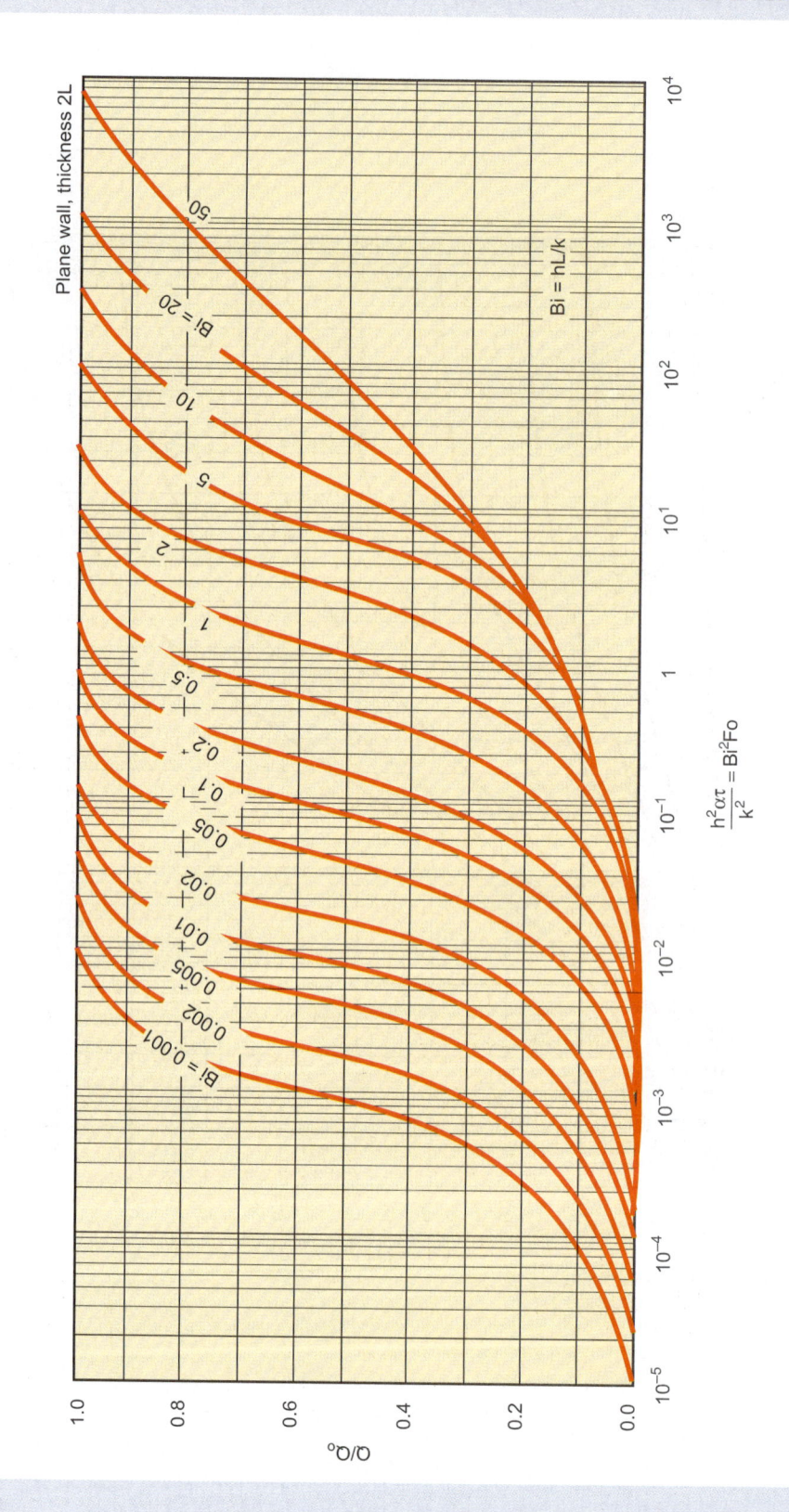

HEAT FLOW: INFINITE PLATE

Plane wall, thickness 2L

$Bi = hL/k$

$\dfrac{h^2 \alpha \tau}{k^2} = Bi^2 Fo$

LONG CYLINDER: TEMPERATURE—TIME HISTORY AT CENTRELINE

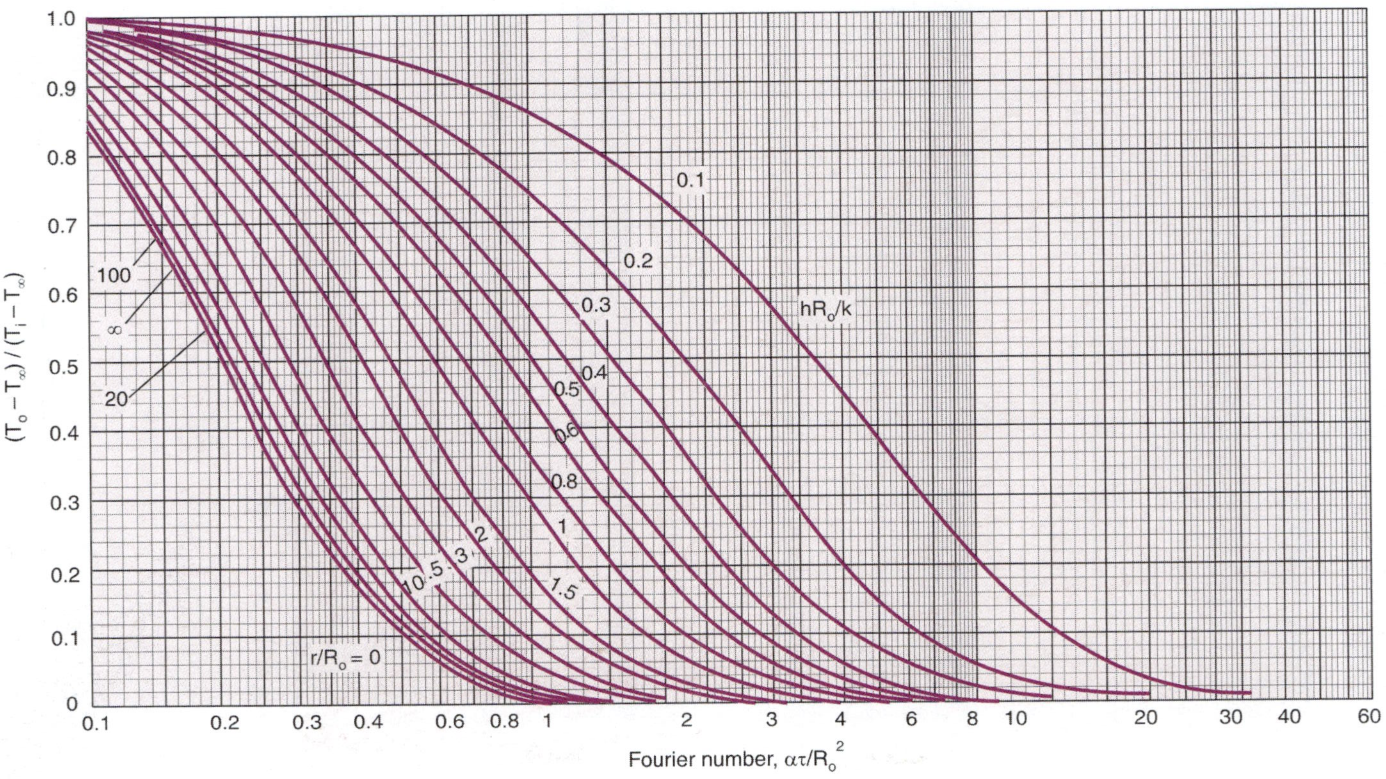

LONG CYLINDER: TEMPERATURE—TIME HISTORY AT ANY RADIUS, r

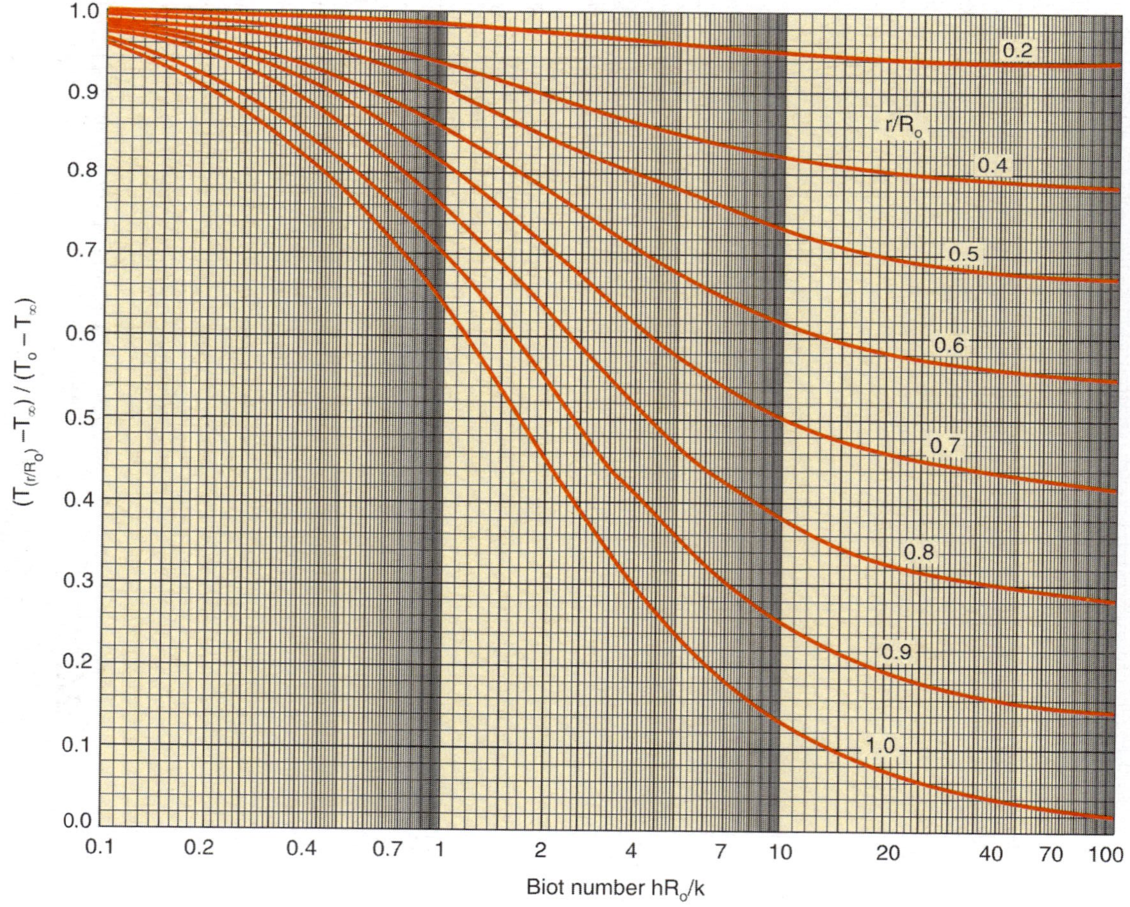

HEAT FLOW: LONG CYLINDER

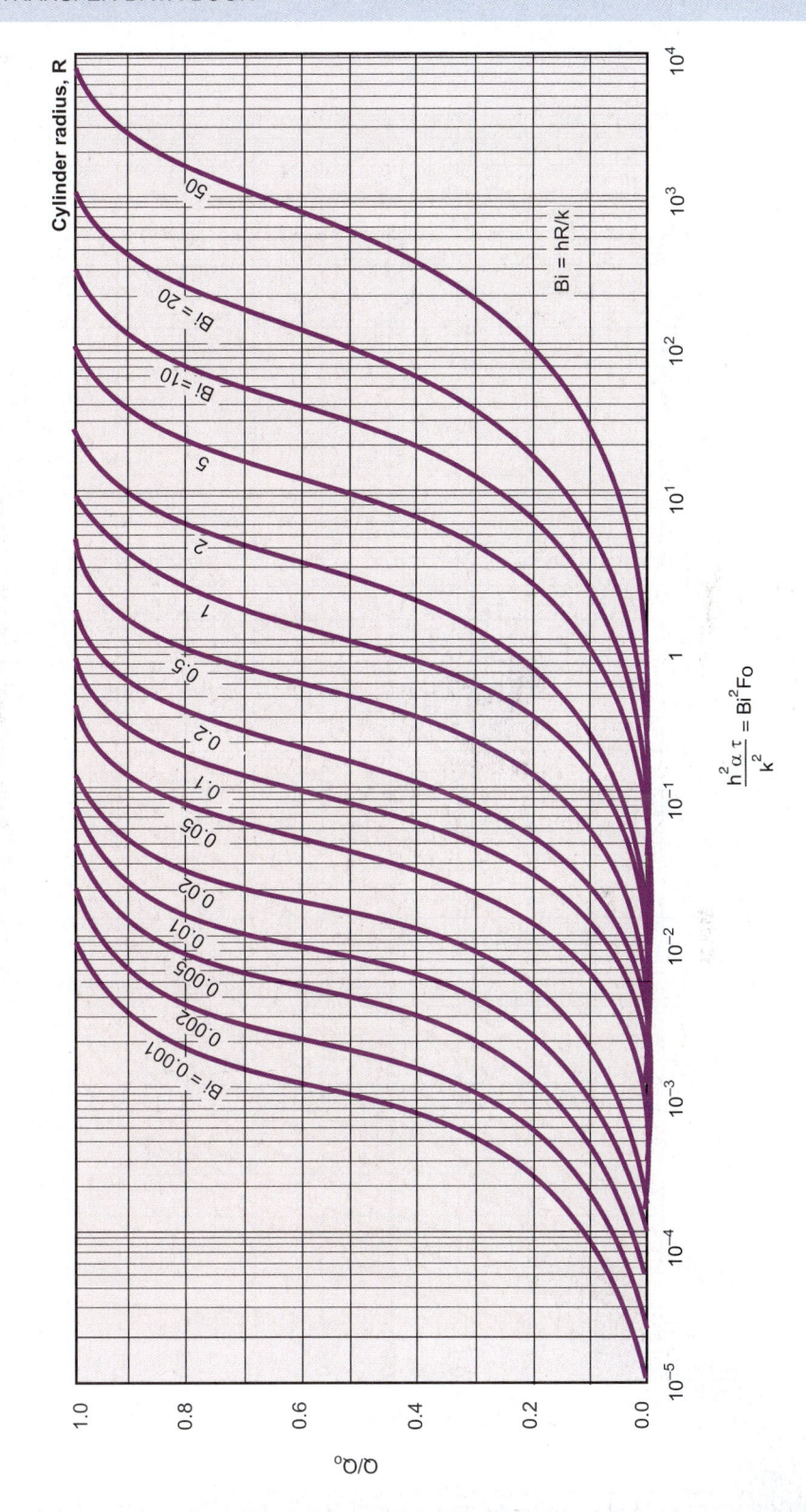

SPHERE: TEMPERATURE—TIME HISTORY AT CENTRE

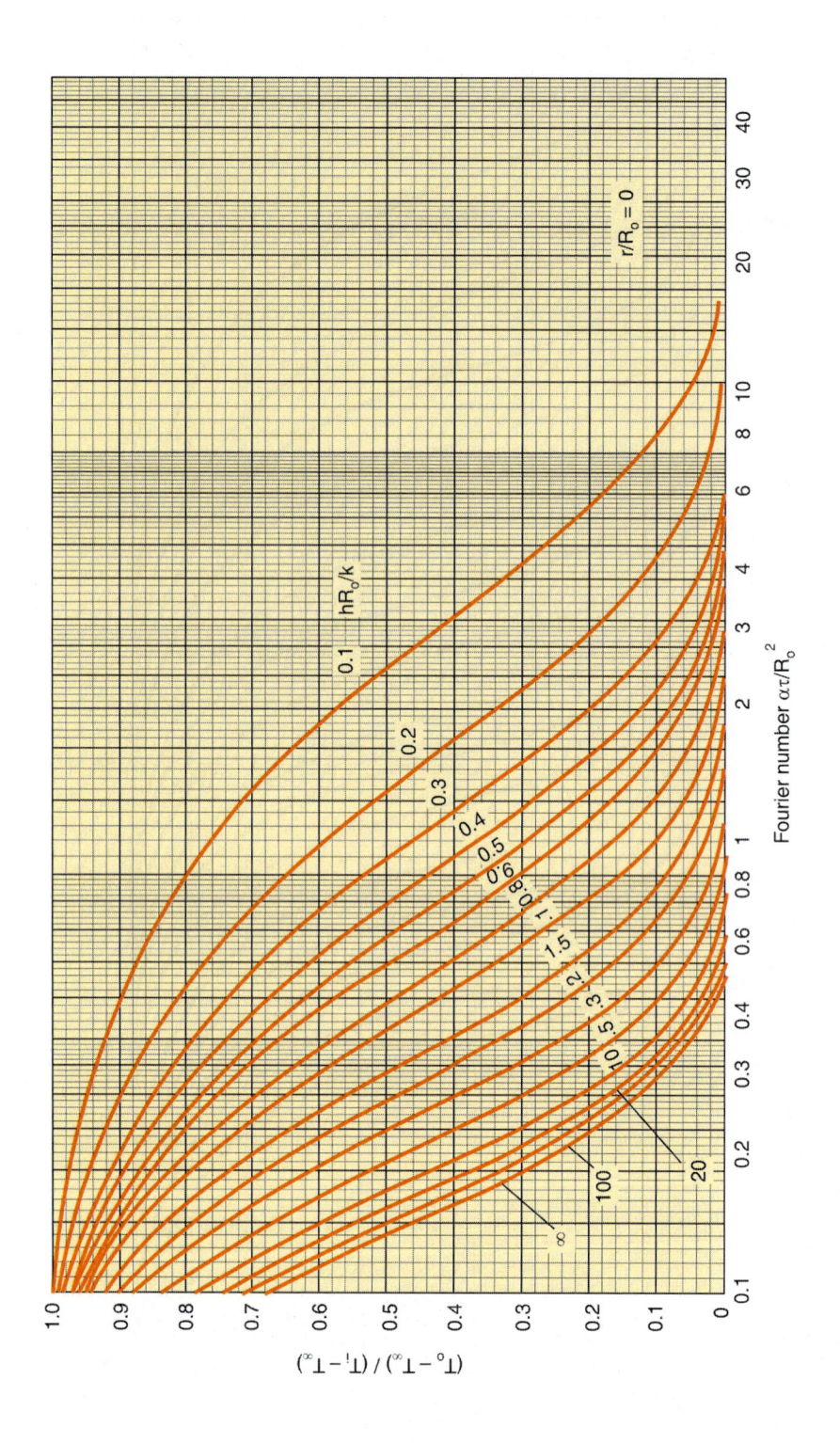

SPHERE: TEMPERATURE—TIME HISTORY AT ANY RADIUS, r

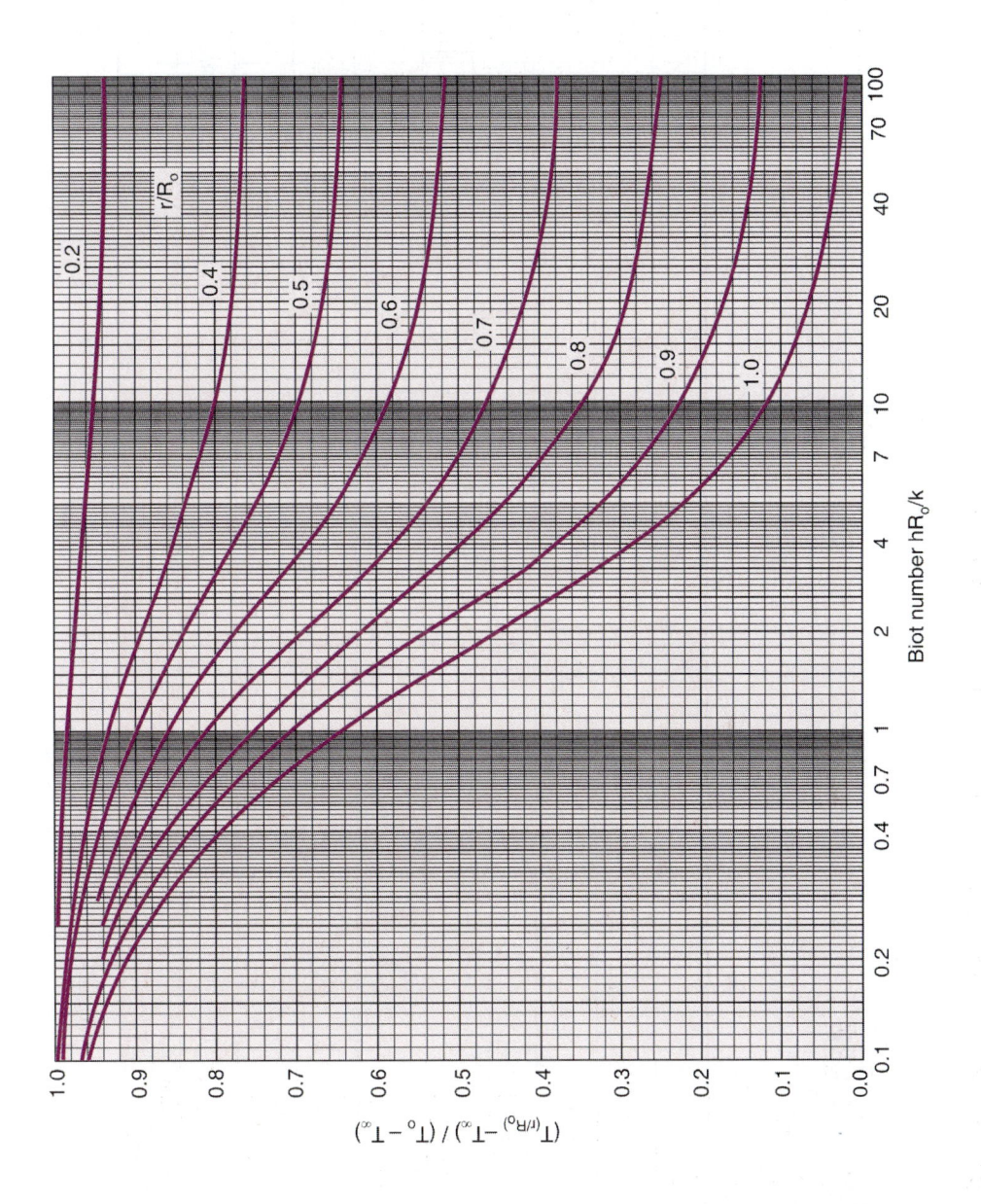

HEAT FLOW: SPHERE

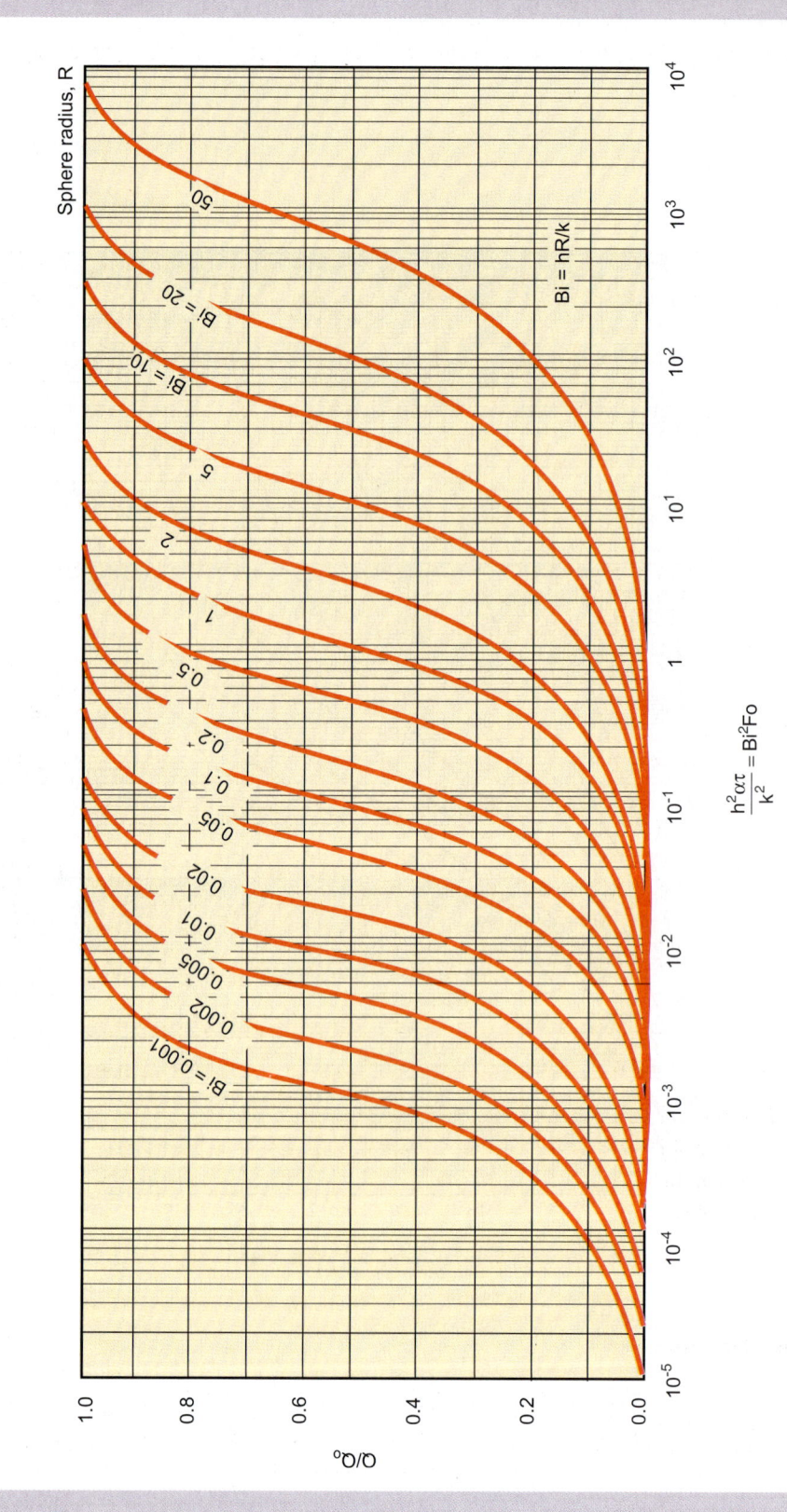

PRODUCT SOLUTIONS FOR TEMPERATURES IN MULTIDIMENSIONAL SYSTEMS

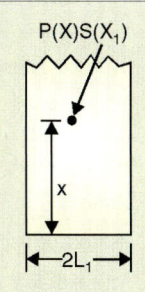

Semi-infinite plate — $P(X)S(X_1)$, width $2L_1$

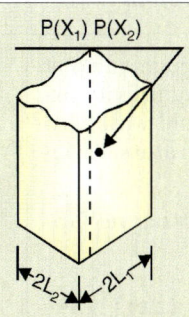

Infinite rectangular bar — $P(X_1) P(X_2)$, $2L_2 \times 2L_1$

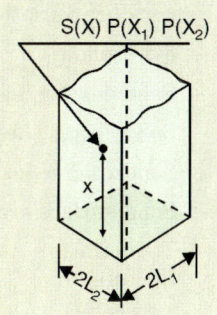

Semi-infinite rectangular bar — $S(X) P(X_1) P(X_2)$, $2L_2 \times 2L_1$

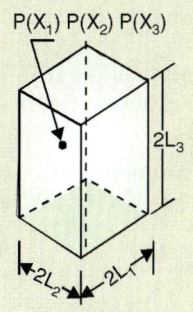

Rectangular parallelepiped — $P(X_1) P(X_2) P(X_3)$, $2L_2 \times 2L_1 \times 2L_3$

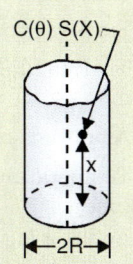

Semi-infinite cylinder — $C(\theta) S(X)$, $2R$

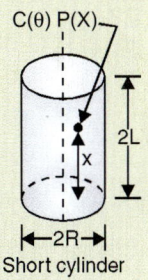

Short cylinder — $C(\theta) P(X)$, $2R \times 2L$

P(X)   Solution for infinite plate
S(X)   Solution for semi-infinite bodies
C(θ)   Solution for infinite cylinder

Intersection of two bodies :

$$\left(\frac{Q}{Q_o}\right)_{Total} = \left(\frac{Q}{Q_o}\right)_1 + \left(\frac{Q}{Q_o}\right)_2 \left[1 - \left(\frac{Q}{Q_o}\right)_1\right]$$

Intersection of three one dimensional systems :

$$\left(\frac{Q}{Q_o}\right)_{Total} = \left(\frac{Q}{Q_o}\right)_1 + \left(\frac{Q}{Q_o}\right)_2 \left[1 - \left(\frac{Q}{Q_o}\right)_1\right] + \left(\frac{Q}{Q_o}\right)_3 \left[1 - \left(\frac{Q}{Q_o}\right)_1\right]\left[1 - \left(\frac{Q}{Q_o}\right)_2\right]$$

PERIODIC HEAT FLOW

Description	Correlation and Validity	Notations
1. Solid with high value of thermal conductivity exposed to fluid with periodic temperature variation. Fluid Temperature Variation: $T_f = T_a \cdot \cos\left(\dfrac{2\pi\tau}{\tau_o}\right)$	$\delta = \tan^{-1}\left(\dfrac{2\pi}{\tau_o} \cdot \dfrac{\rho cV}{hA}\right)$ $\dfrac{T_s}{T_a} = \dfrac{1}{(1 + \tan^2 \delta)^{0.5}}$	T_f — fluid temperature at time τ, K T_a — Amplitude of temperature variation of fluid, K τ — Time, s τ_o — Period of oscillation, s δ — Angle of lag in the temperature variation of the solid with reference to fluid temperature variation
2. Semi infinite solid-surface temperature varied periodically Surface Temperature variation: $T_w = T_a \cos \dfrac{2\pi\tau}{\tau_o}$	$\dfrac{T_{sx}}{T_a} = e^{-x(\pi/\alpha\tau_o)^{0.5}}$ $\dfrac{T_{x\tau}}{T_{sx}} = \cos\left(\dfrac{2\pi\tau}{\tau_o} - x\sqrt{\dfrac{\pi}{\alpha\tau_o}}\right)$ $\delta_x = 0.5\, x(\tau_o/\alpha\pi)^{0.5}$ Depth of penetration at steady state: $x_o = 1.6\,(\pi\alpha\tau_o)^{0.5}$ $\dfrac{q}{A} = kT_a\,(2\pi/\alpha\tau_o)^{0.5} \cdot \sin\left(\dfrac{\pi}{4} - \dfrac{2\pi\tau}{\tau_o}\right)$ $\dfrac{Q}{A} = k \cdot T_a \left(\dfrac{2\tau_o}{\pi\alpha}\right)^{0.5}$	δ_x — δ at x plane ρ — Solid density, m³/kg c — Solid specific heat, J/kgK V — Solid volume, m³ A — Solid surface area, m² T_s — Temperature amplitude of the solid, K $\dfrac{T_s}{T_a}$ — Amplitude ratio T_w — Surface temperature, K α — Thermal diffusivity of solid, m²/s T_{sx} — Amplitude at depth x, K $T_{x\tau}$ — Temperature at x plane at time τ. δ_x — Angle of lag of temperature at depth x, with reference to surface temperature
3. Semi infinite solid surface exposed to convection with periodic temperature variation of fluid.	$T_f = T_a \cos\left(\dfrac{2\pi\tau}{\tau_o}\right)$ $\dfrac{T_{sx}}{T_a} = \dfrac{1}{[1 + 2A^{0.5} + 2A]^{0.5}}$ where $A = (\pi k^2/\alpha\tau_o h^2)$	x_o — Depth upto which the temperature variation penetrates, at steady state. $\dfrac{q}{A}$ — Heat flow rate at time τ, $\dfrac{W}{m^2}$ $\dfrac{Q}{A}$ — Total heat flow during half period, J k — Solid thermal conductivity h — Convective heat transfer coefficient.

FREEZING/MELTING

Description	Correlation	Notations
LIQUID/SOLID SURFACE Suddenly exposed to fluid at T_∞ liquid or solid at melting temperature T_{fr}. Flat surface Refer page 78	$x^* = [2\tau^* + 1]^{0.5} - 1$ $\tau^* = x^* + 0.5\, x^{*2}$ where $x^* = \dfrac{x \cdot h_\infty}{k_1}$ melting $x^* = \dfrac{x \cdot h_\infty}{k_s}$ freezing $\tau^* = \tau\, \dfrac{h_\infty^2\, (T_\infty - T_{fr})}{k_1 \cdot \rho_s \cdot h_{sf}}$ – melting $\tau^* = \tau\, \dfrac{h_\infty^2\, (T_{fr} - T_\infty)}{k_s \cdot \rho_1 \cdot h_{sf}}$ – freezing	x — thickness melted/frozen from surface upto time τ, m x^* — dimensionless distance parameter τ^* — dimensionless time parameter k_s, k_1 — thermal conductivity of solid, liquid ρ_s, ρ_1 — density of solid, liquid h_{sf} — enthalpy of melting/freezing, J/kg h_∞ — convective heat transfer coefficient at surface, W/m²K T_∞ — free stream temperatue of fluid flowing over surface T_{fr} — melting/freezing temperature, τ — time, s

FREEZING/MELTING: FLAT SURFACE

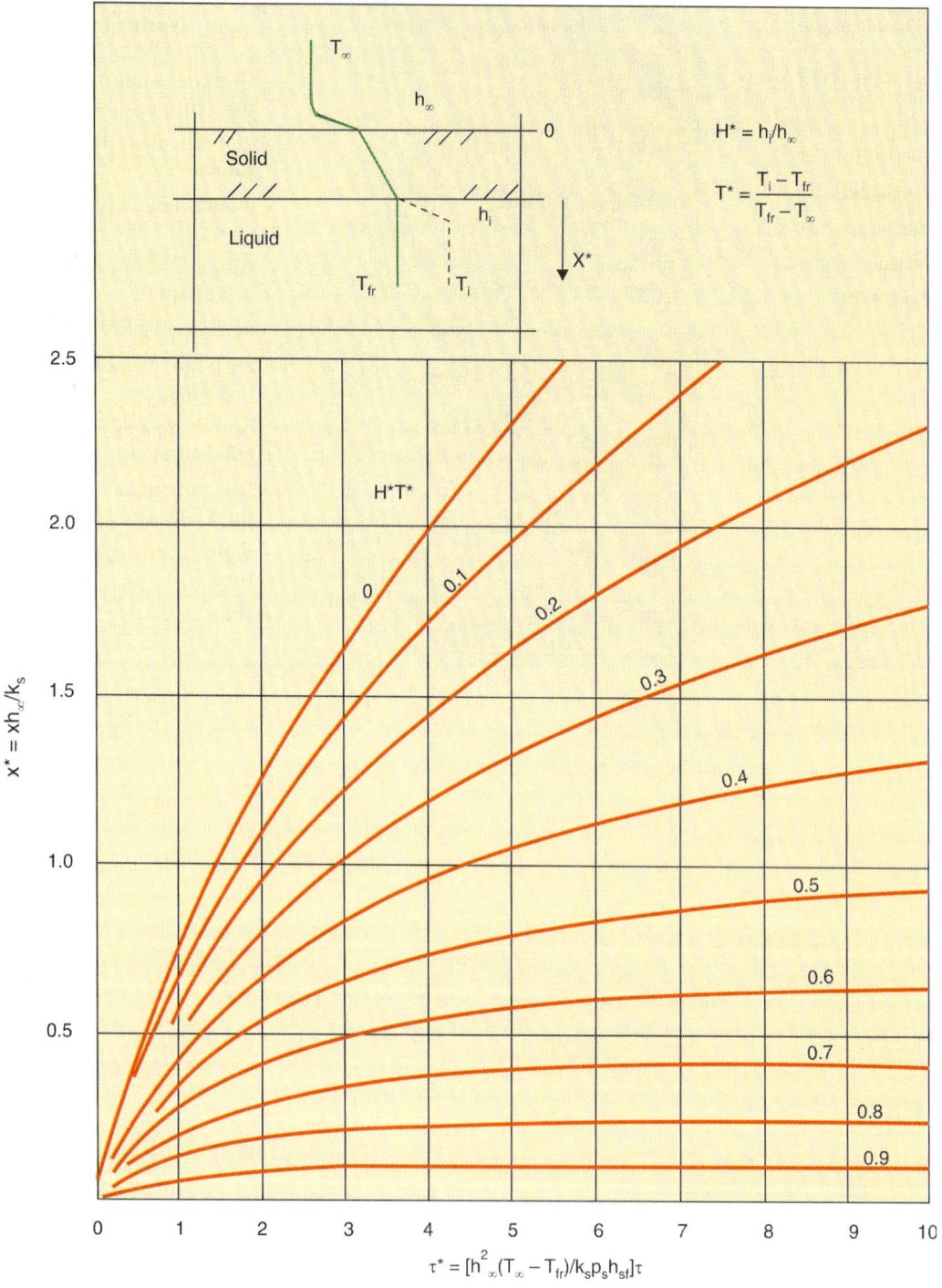

$H^* = h_i/h_\infty$

$T^* = \dfrac{T_i - T_{fr}}{T_{fr} - T_\infty}$

$x^* = xh_\infty/k_s$

$\tau^* = [h_\infty^2 (T_\infty - T_{fr})/k_s \rho_s h_{sf}]\tau$

Variation of the Freezing Thickness x^* with Dimensionless Time τ^* During Slab Formation of the Solid Phase

FREEZING/MELTING: FLAT SURFACE *(Contd.)*

Description	Correlation	Notations
Freezing of liquid at a temperature higher than freezing point, during slab formation	$\tau^* = \left[\dfrac{1}{H^*T^*}\right]^2$ $\ln\dfrac{1-H^*T^*}{1-H^*T^*(1+x^*)} - \dfrac{x^*}{H^*T^*}$ where $T^* = \dfrac{T_i - T_{fr}}{T_{fr} - T_\infty}$, $x^* = \dfrac{xh_\infty}{k_s}$ $\tau^* = \tau\dfrac{h_\infty^2(T_{fr}-T_\infty)}{k_s \cdot \rho_s \cdot h_{sf}}$, $H^* = \dfrac{h_1}{h_\infty}$	h_1 — convection coefficient at the liquid solid interface T_1 — Temperature of liquid x — thickness from surface frozen upto time τ, m
Freezing inside a tube surface with convection outside the tube surface OR Freezing outside the tube surface with convection inside the tube surface	Freezing inside : refer Page 80 $\tau^* = 0.5\, r^{*2} \ln r^*$ $+ \left(\dfrac{1}{2R^*} + 0.25\right)(1 - r^{*2})$ Freezing outside : refer Page 81 $\tau^* = 0.5\, r^{*2} \cdot \ln r^*$ $+ \left(\dfrac{1}{2R^*} - 0.25\right)(r^{*2} - 1)$ $\tau^* = \tau\dfrac{(T_{fr}-T_\infty)k_s}{\rho_s\, h_{sf} r_o^2}$ $r^* = r/r_o,\ R^* = \dfrac{h_\infty \cdot r_o}{k_s}$	r — radius at the (freezing) solid/liquid interface at time τ, m r_o — tube outside radius, m

FREEZING: INSIDE TUBE

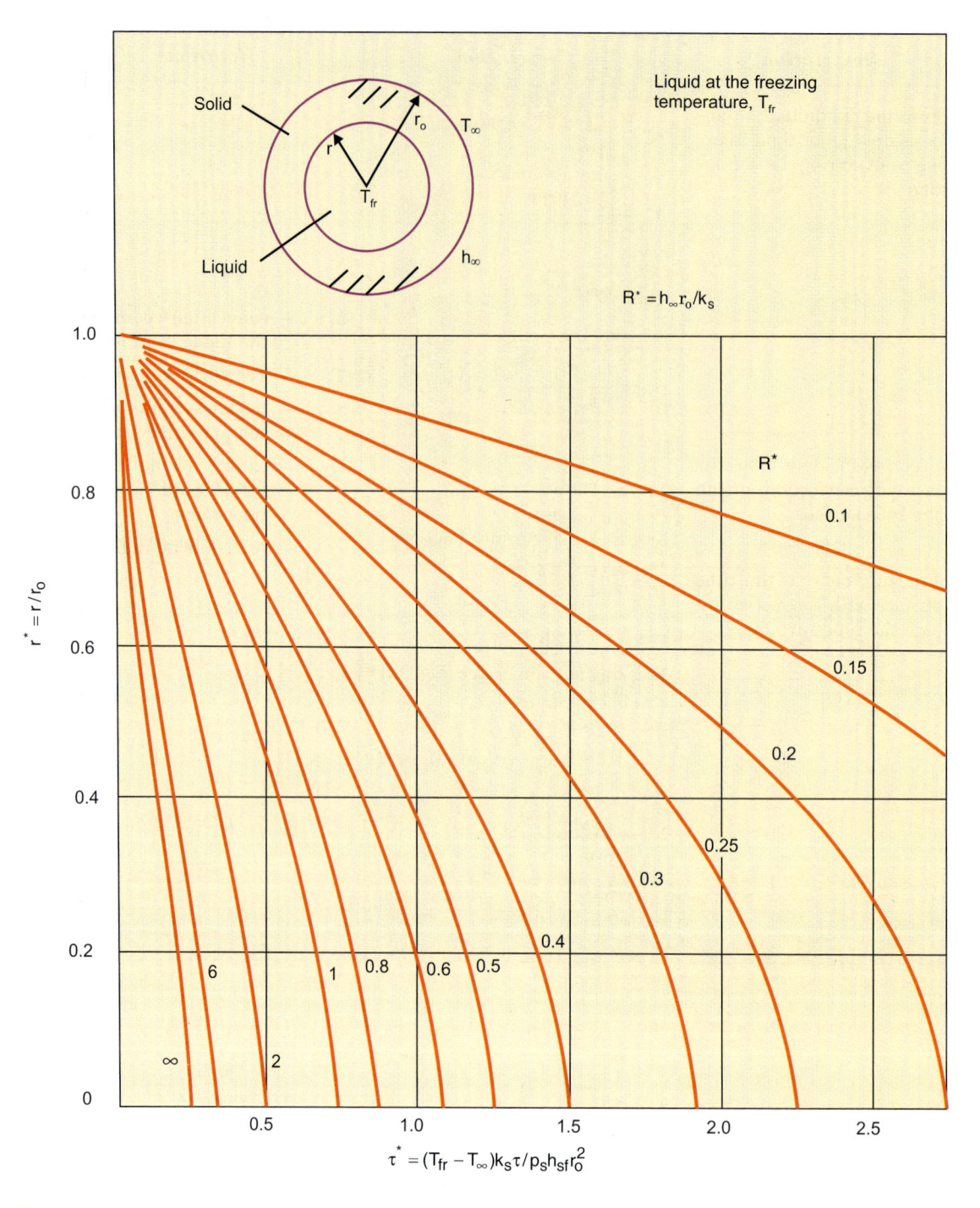

Variation of r^* with Dimensionless Time τ^* for Freezing inside a Tube of Negligible Thermal Resistance Having Convection on the Outside and Ambient Sink Temperature T_∞.

$R^* = h_\infty r_o / k_s$

$r^* = r/r_o$

$\tau^* = (T_{fr} - T_\infty) k_s \tau / \rho_s h_{sf} r_o^2$

FREEZING: OUTSIDE TUBE

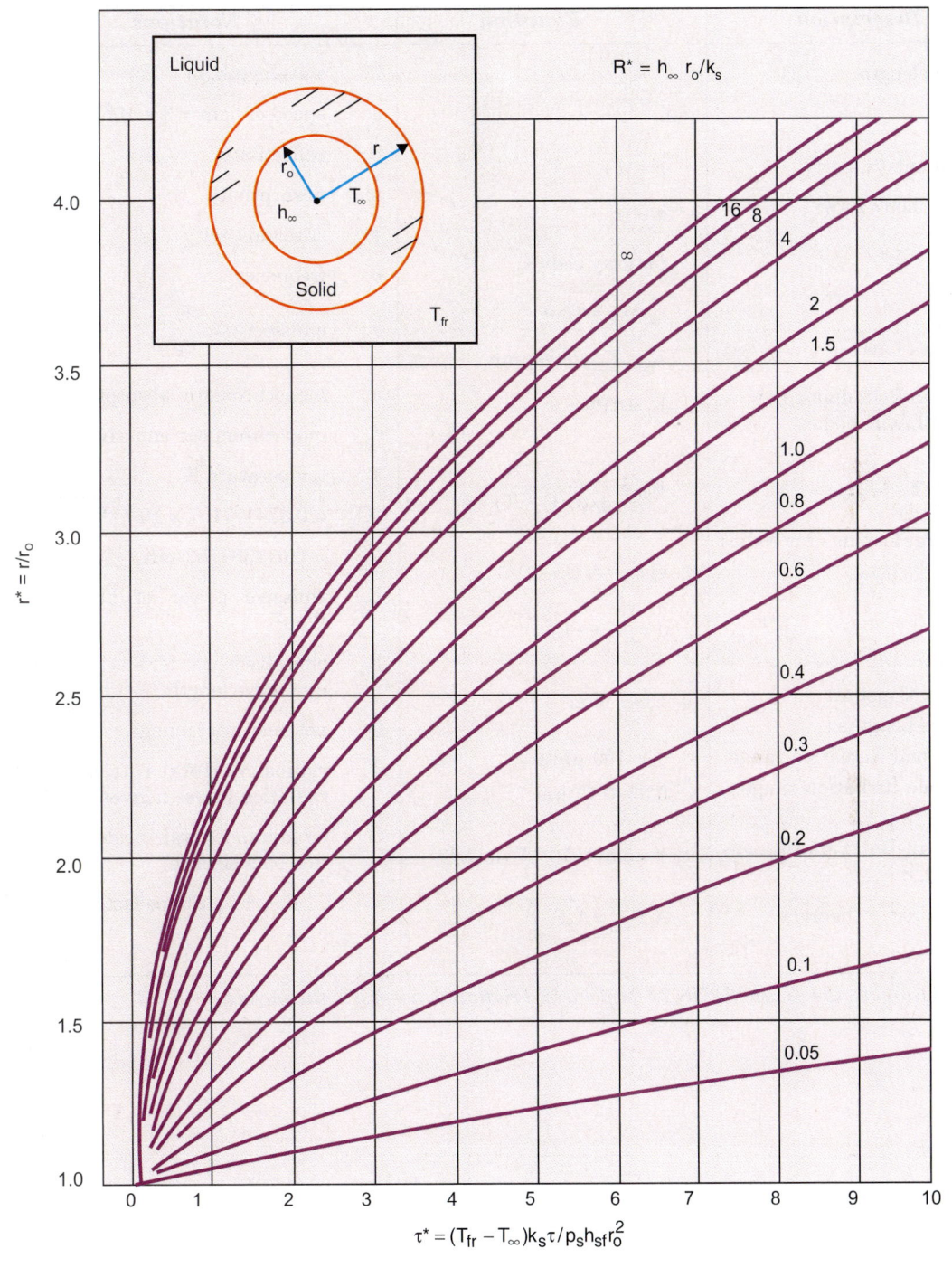

Variation of r^* with Nondimensional Time τ^* for Freezing on the Outside of Tube of Negligible Thermal Resistance Having Convection on the Inside and Sink Temperature T_∞.

Axes: $r^* = r/r_o$ versus $\tau^* = (T_{fr} - T_\infty)k_s\tau/\rho_s h_{sf} r_o^2$

$R^* = h_\infty r_o/k_s$

Curves labelled: ∞, 16, 8, 4, 2, 1.5, 1.0, 0.8, 0.6, 0.4, 0.3, 0.2, 0.1, 0.05

RADIATION

Description	Equation	Notations
Wavelength	$\lambda = \dfrac{c}{\upsilon}$	λ wavelength, m c speed of light = 3×10^8 m/s
Radiant Property	$\rho + \alpha + \tau = 1$	ρ reflectivity
Kirchhoff's Law	$\varepsilon_\lambda = \alpha_\lambda$ for gray bodies $\varepsilon_\lambda = \alpha_\lambda = \varepsilon = \alpha$	α absorptivity τ transmissivity υ frequency ε emissivity = $\dfrac{E}{E_b}$
Wien's Law	$\lambda_{max} T = 2898$ μmK	
Stefan-Boltzmann Law for BLACK bodies	$E_b = \sigma T^4$	α_λ monochromatic absorptivity ε_λ monochromatic emissivity
Planck's Law	$E_{b\lambda} = \dfrac{C_1 \lambda^{-5}}{\exp.(C_2/\lambda T) - 1}$	T temperature K *C_1 = $0.374177107 \times 10^{-15}$ W/m^2 C_2 = 0.014387752 mK
Energy leaving gray bodies	$Q = \dfrac{\varepsilon}{\rho} A(E_b - J)$ $J = \rho G + \varepsilon E_b$	E_b emissive power of black body, W/m^2 $E_{b\lambda}$ emissive power for black body per unit wave length
Solar Constant	1395 W/m^2	Q net heat interchange
Wave Lengths : Thermal Radiation range Visible Radiation range	 0.1–100 μm 0.35–0.75 μm	J radiosity (total rate at which radiation leaves a given surface)
Stefan-Boltzmann constant, σ	$\sigma = 5.67 \times 10^{-8}$ W/m^{2}K^4	G irradiation (total incident radiation on a surface)
Reciprocity theorem	$A_1 F_{1-2} = A_2 F_{2-1}$	F_{1-2}, F_{2-1}, Shape factors

*The constant C_1 is to be divided by 2π for polarised radiation normal to the surface.

RATIO OF BLACK BODY RADIANT ENERGY UPTO THE WAVE LENGTH λ, TO TOTAL EMISSIVE POWER OF BLACK BODY AT THE SAME TEMPERATURE BLACK BODY RADIATION FUNCTION, $E_{b0-\lambda}/\sigma T^4$

λT, μm-K	$E_{b0-\lambda}/\sigma T^4$	λT, μm-K	$E_{b0-\lambda}/\sigma T^4$
800	0.000016	6800	0.796129
1000	0.000321	7000	0.808109
1200	0.002134	7200	0.819217
1400	0.007790	7400	0.829527
1600	0.019718	7600	0.839102
1800	0.039341	7800	0.848005
2000	0.066728	8000	0.856288
2200	0.100888	8500	0.874608
2400	0.140256	9000	0.890090
2600	0.183120	9500	0.903085
2800	0.227897	10000	0.914199
3000	0.273232	10500	0.923710
3200	0.318102	11000	0.931890
3400	0.361735	11500	0.939959
3600	0.403607	12000	0.945098
3800	0.443382	13000	0.955139
4000	0.480877	14000	0.962898
4200	0.516014	15000	0.969981
4400	0.548796	16000	0.973814
4600	0.579280	18000	0.980860
4800	0.607559	20000	0.985602
5000	0.633747	25000	0.992215
5200	0.658970	30000	0.995340
5400	0.680360	40000	0.997967
5600	0.701046	50000	0.998953
5800	0.720158	75000	0.999713
6000	0.737818	100000	0.999905
6200	0.754140		
6400	0.769234		
6600	0.783199		

RADIANT HEAT EXCHANGE BETWEEN SURFACES

General Equations	$Q_{ij} = [E_{bi} - E_{bj}] F_{ij} A_i$		Q_{ij}	Net heat exchange between i and j
	$F_{ij} A_i = F_{ji} A_j$	E_{bj} emissive power of body j	E_{bi}	emissive power of body i
			F_{ij}	shape factor of surface i with respect to j

Geometry	Shape factor F_{ij} for black bodies	For gray bodies $A_i F_{ij}$
Equal and parallel squares, rectangles and discs	Refer pages 91 to 93	
Adjacent rectangles at right angles	Refer pages 94 to 95	
Non intersecting perpendicular squares	Refer page 96	
Offset non intersecting perpendicular squares	Refer page 97	$\dfrac{1}{\dfrac{1-\varepsilon_i}{A_i \varepsilon_i} + \dfrac{1}{A_i F_{ij}} + \dfrac{1-\varepsilon_j}{A_j \varepsilon_j}}$
Parallel discs of different size	Refer page 98	
Opening in walls of thickness L	Refer page 99	
Surface element and a parallel rectangular surface	Refer page 100	
Surface element and a perpendicular rectangular surface	Refer page 101	
Differential spherical surface and a rectangular surface	Refer page 101	
A plane with parallel rows of tubes	Refer page 102 and 103	
Short concentric cylinders	Refer page 104	
Parallel infinite planes	1	
Totally enclosed small body i [small compared with a large enclosing body j]	1	
Totally enclosed large convex body	1	
Concentric infinite long cylinders,	1	
Concentric spheres	1	

RELATION BETWEEN SHAPE FACTORS

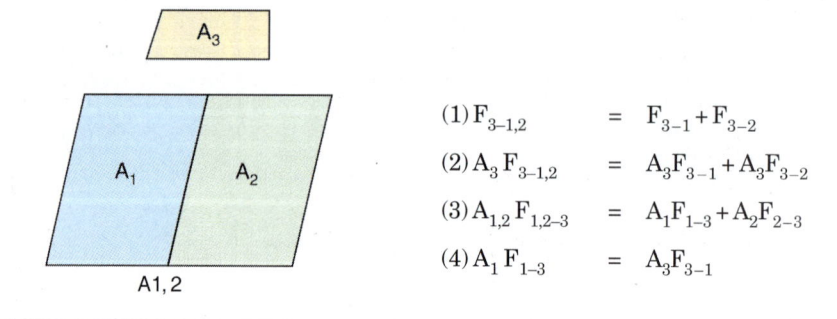

(1) $F_{3-1,2} = F_{3-1} + F_{3-2}$

(2) $A_3 F_{3-1,2} = A_3 F_{3-1} + A_3 F_{3-2}$

(3) $A_{1,2} F_{1,2-3} = A_1 F_{1-3} + A_2 F_{2-3}$

(4) $A_1 F_{1-3} = A_3 F_{3-1}$

EMISSIVITY ε_n OF RADIATION IN THE NORMAL DIRECTION TO THE SURFACE AND ε_h OF THE TOTAL HEMISPHERICAL EMISSIVITY FOR VARIOUS SURFACES

Surface	Temperature t °C	ε_n	ε_h
Aluminium, bright rolled	170	0.039	0.049
Aluminium, bright rolled	500	0.050	—
Aluminium paint	100	0.20–0.40	—
Bakelite lacquer	80	0.935	—
Bismuth, bright	80	0.340	0.366
Brick, mortar, plaster	20	0.930	—
Chrome, polished	150	0.058	0.071
Clay, fired	70	0.910	0.860
Copper, polished	20	0.030	—
Copper, lightly oxidised	20	0.037	—
Copper, scraped	20	0.070	—
Copper, black oxidised	20	0.780	—
Copper, oxidised	131	0.760	0.725
Corrundum, emery rough	80	0.855	0.840
Enamel, lacquer	20	0.85–0.95	—
Glass	90	0.940	0.876
Gold, polished	130	0.018	—
Gold, polished	400	0.022	—
Ice, smooth and water	0	0.966	0.918
Ice, rough crystals	0	0.985	—
Iron, bright etched	150	0.128	0.158
Iron bright abrased	20	0.240	—
Iron, red rusted	20	0.610	—
Iron, hot rolled	20	0.770	—
Iron, hot rolled	130	0.600	—
Iron, hot cast	100	0.800	—
Iron, heavity rusted	0	0.850	—
Iron, heat resistant, oxidised	80	0.613	—
Iron, heat resistant, oxidised	200	0.630	—

(Contd.)

Surface	Temperature t °C	ε_n	ε_h
Lacquer, black matte	80	0.970	—
Lacquer, white	100	0.925	—
Lead, gray oxidised	20	0.280	—
Manganin, bright rolled	118	0.048	0.057
Nickel, bright matte	100	0.041	0.046
Nickel, Polished	100	0.045	0.053
Paper	95	0.920	0.890
Porcelain	20	0.92–0.94	—
Red Lead	100	0.930	—
Silver, pure, polished	200–600	0.02–0.03	—
Silumin, cast polished	150	0.186	—
Tar paper	20	0.930	—
Water glass	20	0.960	—
Wood, beach	70	0.935	0.910
Zinc, gray oxidised	20	0.23–0.28	—

*Where ε_h is not specified take:
For bright metal surfaces, $\varepsilon_h = 1.2\, \varepsilon_n$
For other smooth surfaces = $\varepsilon_h = 0.95\, \varepsilon_n$
For rough surfaces $\varepsilon_h = 0.98\, \varepsilon_n$.

EMISSIVITY ALONG NORMAL DIRECTION ε_n AND TOTAL HEMISPHERICAL EMISSIVITY ε_h FOR VARIOUS NON METALLIC SURFACES

Surface	Temperature °C	ε_h	ε_n
Aluminium, oxide	327	—	0.69
Aluminium, oxide	727	—	0.55
Aluminium oxide	1227	—	0.41
Alumina brick	527	—	0.40
Alumina brick	727	—	0.33
Alumina brick	1127	—	0.28
Alumina brick	1327	—	0.33
Asphalt pavement	27	0.85–0.93	—
Brick, red	27	0.93–0.96	—
Cloth	27	0.75–0.90	—
Concrete	27	0.88–0.93	—
Glass, window	27	0.90–0.95	—
Gypsum or plaster board	27	0.90–0.92	—
Ice	0	0.95–0.98	—
Kaolin insulating brick	527	—	0.70
Kaolin insulating brick	927	—	0.57
Kaolin insulating brick	1127	—	0.47
Kaolin insulating brick	1327	—	0.53
Magnesia brick	527	—	0.45
Magnesia brick	727	—	0.36
Magnesia brick	1127	—	0.31
Magnesia brick	1327	—	0.40
Paints, Black	27	0.98	—
Paints, white acrylic	27	0.90	—
Paints white, zinc oxide	27	0.92	—
Paper, white	27	0.92–0.97	—
Pyrex	27	—	0.82
Pyrex	327	—	0.80
Pyrex	727	—	0.71
Pyrex	927	—	0.62

(Contd.)

Surface	Temperature °C	ε_h	ε_n
Pyroceram	27	—	0.85
Pyroceram	327	—	0.78
Pyroceram	727	—	0.69
Pyroceram	1227	—	0.57
Rocks	27	0.88–0.95	—
Sand	27	0.90	—
Silicon carbide	327	—	0.87
Silicon carbide	727	—	0.87
Silicon carbide	1227	—	0.85
Skin	27	0.95	—
Snow	0	0.82–0.90	—
Soil	27	0.93–0.96	—
Teflon	27	0.85	—
Teflon	127	0.87	—
Teflon	227	0.92	—
Vegetation	27	0.92–0.96	—
Water	27	0.96	—
Wood	27	0.82–0.92	—

SOLAR RADIATIVE PROPERTIES FOR SELECTED MATERIALS

Material	Absorptivity α	Emissivity at 300 K ε	Transmissivity τ	α/ε
Aluminium, polished	0.09	0.03	—	3.0
Aluminium, anodised	0.14	0.84	—	0.17
Aluminium coated over quartz	0.11	0.37	—	0.30
Aluminium foil	0.15	0.05	—	3.0
Brick, red	0.63	0.93	—	0.68
Concrete	0.60	0.88	—	0.68
Galvanised sheet metal, clean	0.65	0.13	—	5.0
Galvanised oxidised, weathered	0.80	0.28	—	2.9
Glass, 3.2 mm, thickness :				
Float or tempered	—	—	0.78	—
Low iron oxide type	—	—	0.88	—
Metals, plated :				
Black sulfide	0.92	0.10	—	9.2
Black cobalt oxide	0.93	0.30	—	3.1
Black Nickel oxide	0.92	0.08	—	11.0
Black chrome	0.87	0.09	—	9.7
Mylar, 0.13 mm thickness	—	—	0.87	—
Paints : Black	0.98	0.98	—	1.0
White acrylic	0.26	0.90	—	0.29
White zinc oxide	0.16	0.92	—	0.17
Plexi glass, 3.2 mm thickness	—	—	0.90	—
Snow, fresh, fine particles	0.13	0.82	—	0.16
Snow ice granules	0.33	0.89	—	0.37
Tedlar, 0.10 mm	—	—	0.92	—
Teflon, 0.13 mm	—	—	0.92	—

VARIATION OF TOTAL REFLECTIVITY AND ABSORPTIVITY FOR INCIDENT BLACK RADIATION WITH TEMPERATURE

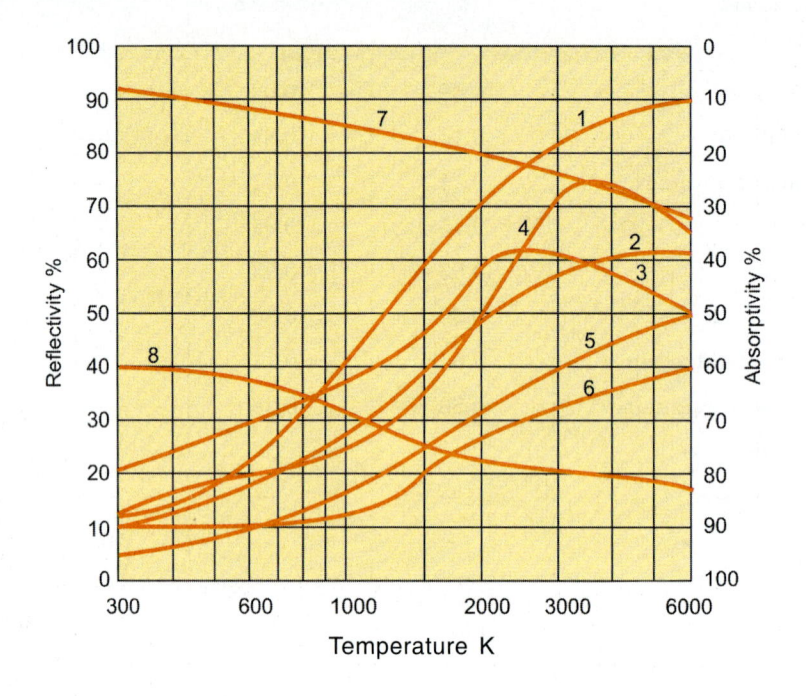

1. White fire clay
2. Asbestos
3. Cork
4. Wood
5. Porcelain
6. Concrete
7. Aluminium
8. Graphite

SHAPE FACTORS FOR EQUAL AND PARALLEL SQUARES, RECTANGLES AND DISCS

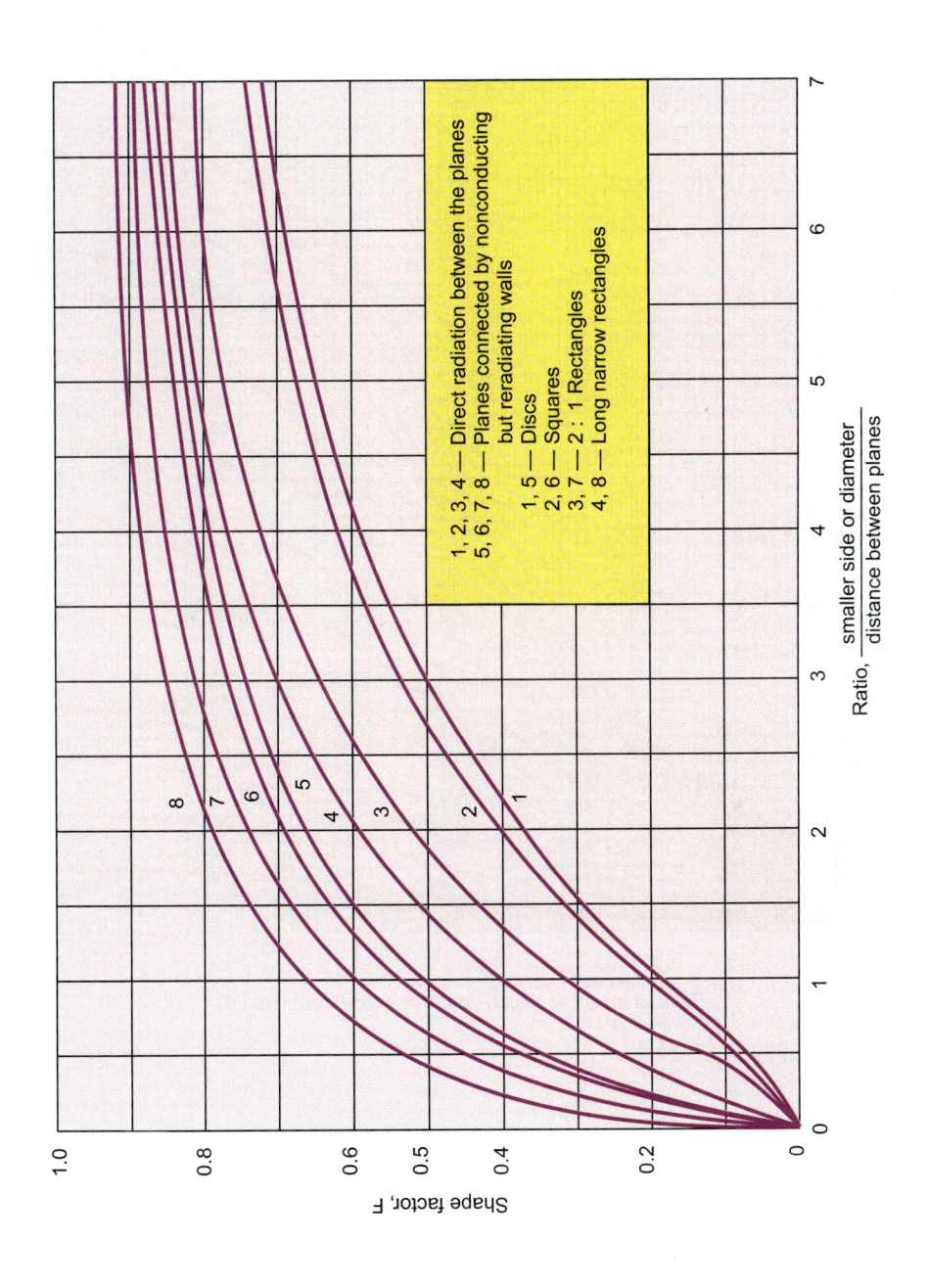

1, 2, 3, 4 — Direct radiation between the planes
5, 6, 7, 8 — Planes connected by nonconducting but reradiating walls

1, 5 — Discs
2, 6 — Squares
3, 7 — 2 : 1 Rectangles
4, 8 — Long narrow rectangles

Ratio, $\dfrac{\text{smaller side or diameter}}{\text{distance between planes}}$

Shape factor, F

SHAPE FACTOR FOR EQUAL PARALLEL RECTANGLES IN OPPOSITE LOCATION

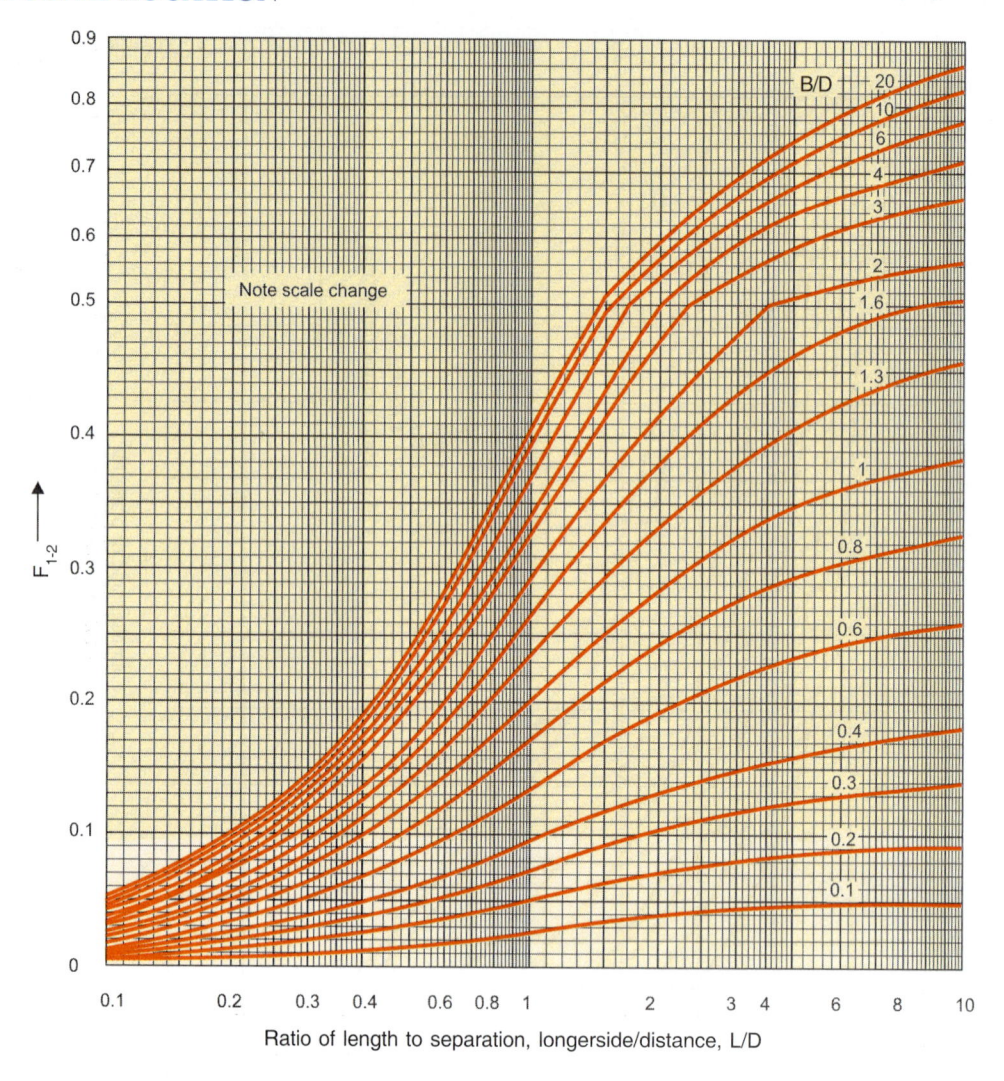

(See next page for notations)

SHAPE FACTORS F_{1-2} : EQUAL PARALLEL RECTANGLES IN OPPOSITE LOCATION

$Y_{B/D}$ \ $X_{L/D}$	0.1	0.2	0.4	0.6	1.0	2.0	4.0	10.0	∞
0.1	.00316	.00626	.01207	.01715	.02492	.03514	.04209	0.4671	.04988
0.2	.00626	.01240	.02392	.03398	.04941	.06971	.08353	.09270	.09901
0.4	.01207	.02391	.04614	.06560	.09554	.13513	.16219	.18021	.19258
0.6	.01715	.03398	.06560	.09336	.13627	.19342	.23171	.25896	.27698
1.0	.02492	.04941	.09554	.13627	.19982	.28588	.34596	.38638	.41421
2.0	.03514	.06971	.13513	.19341	.28588	.41525	.50899	.57361	.61803
4.0	.04210	.08353	.16219	.23271	.34596	.50899	.63204	.71933	.78078
6.0	.04463	.08859	.17209	.24712	.36813	.54421	.67954	.77741	.84713
10.0	.04671	.09272	.18021	.25896	.38638	.57338	.71933	.82699	.90499
20.0	.04829	.09586	.18638	.26795	.40026	.59563	.74990	.86563	.95125
∞	.04988	.09902	.19258	.27698	.41421	.61803	.78078	.90499	1.00000

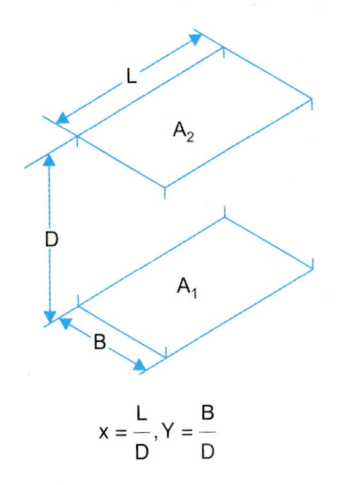

$$x = \frac{L}{D}, Y = \frac{B}{D}$$

(See previous page for chart)

SHAPE FACTORS F_{1-2}—PERPENDICULAR RECTANGLES
(See next page for notations)

SHAPE FACTORS F_{1-2}—PERPENDICULAR RECTANGLES

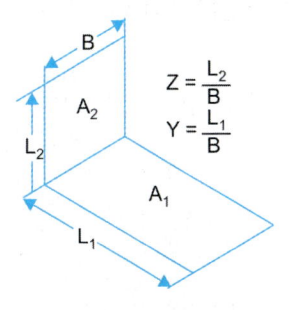

$Z = \dfrac{L_2}{B}$

$Y = \dfrac{L_1}{B}$

$Y \rightarrow$ L_1/B Z $L_2/B \downarrow$	0.02	0.05	0.1	0.2	0.4	0.6	1.0	2.0	4.0	6.0	10.0	20.0
.05	.39908	.28738	.18601	.10584	.05606	.03799	.02304	.01158	.00580	.00388	.00232	.00116
0.1	.44375	.37202	.28189	.18108	.10215	.07048	.04326	.02188	.01097	.00732	.00439	.00220
0.2	.46615	.42337	.36216	.27104	.17147	.12295	.07744	.03971	.02000	.01335	.00802	.00401
0.4	.47725	.44852	.40859	.34295	.25032	.19206	.12770	.06757	.03434	.02296	.01380	.00691
0.6	.47943	.45587	.42290	.36884	.28809	.23147	.16138	.08829	.04536	.03040	.01829	.00916
1.0	.48138	.46073	.43251	.38719	.31924	.26896	.20004	.11643	.06131	.04130	.02492	.01249
2.0	.48239	.46327	.43756	.39711	.33784	.29429	.23285	.14930	.08365	.05731	.03491	.01757
4.0	.48267	.46397	.43898	.39994	.34339	.30238	.24522	.16731	.10136	.07184	.04484	.02285
6.0	.48273	.46411	.43925	.40048	.34447	.30399	.24783	.17193	.10776	.07822	.04998	.02587
10.0	.48275	.46418	.43939	.40076	.34503	.30482	.24921	.17455	.11210	.08331	.05502	.02938
20.0	.48276	.46421	.43943	.40089	.34528	.30518	.24980	.17571	.11427	.08624	.05876	.03302
∞	.48277	.46422	.43947	.40092	.34535	.30530	.25000	.17611	.11505	.08738	.06053	.03578

Note carefully positions of A_1 and A_2

SHAPE FACTOR—NON INTERSECTING PERPENDICULAR EQUAL SQUARES

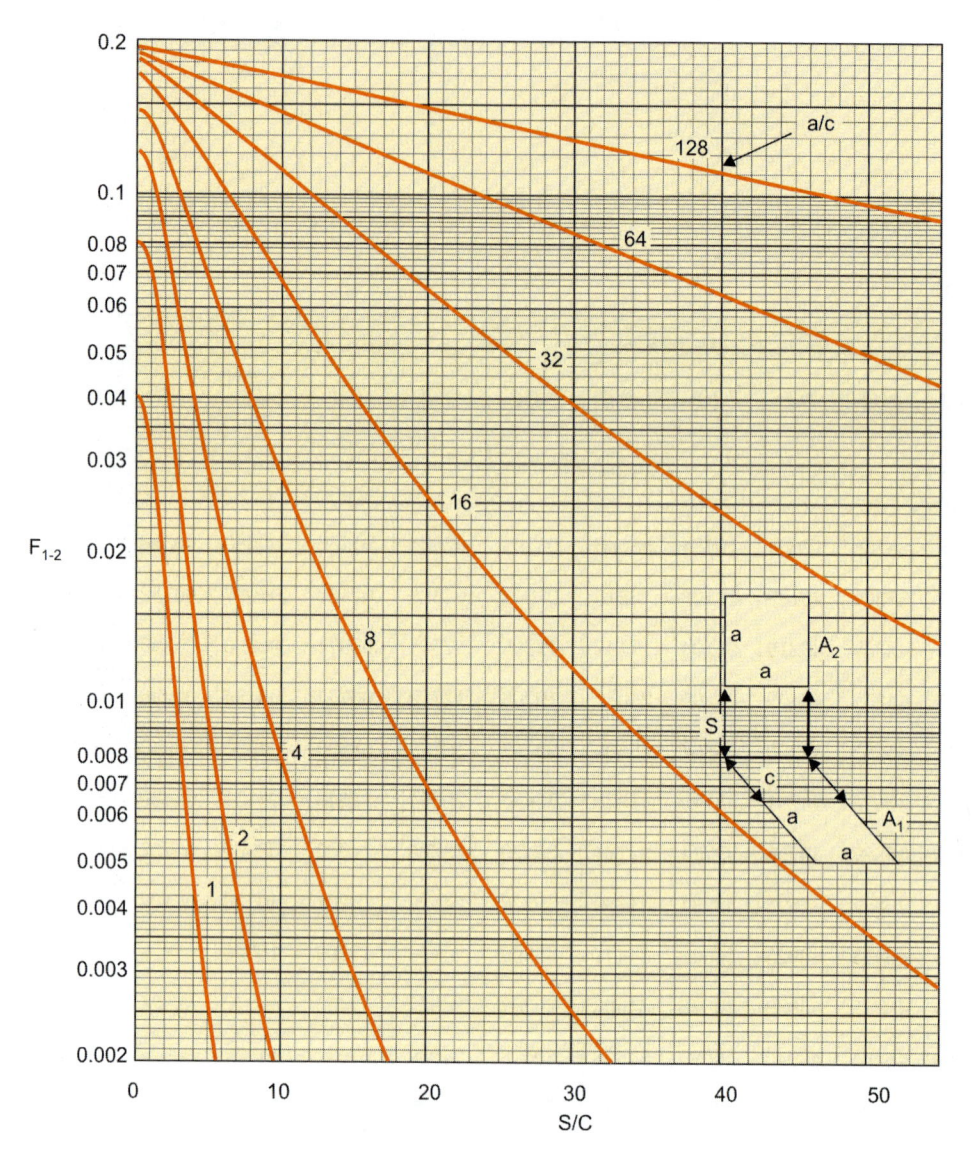

Note positions of A_1 and A_2 carefully.

SHAPE FACTOR—OFFSET NON INTERSECTING PERPENDICULAR SQUARES

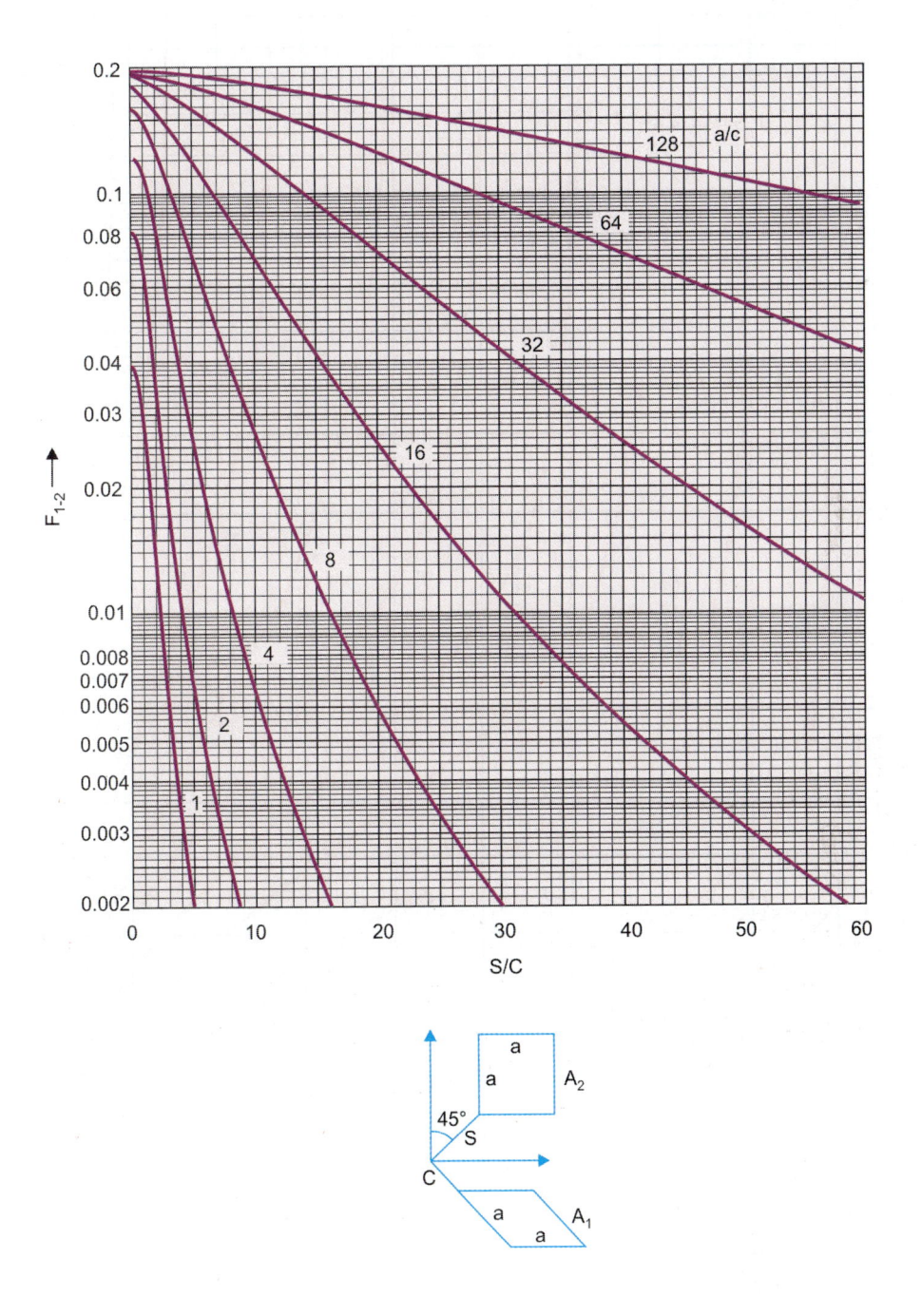

Note positions of A_1 and A_2 carefully.

SHAPE FACTOR—COAXIAL DISCS OF DIFFERENT RADII

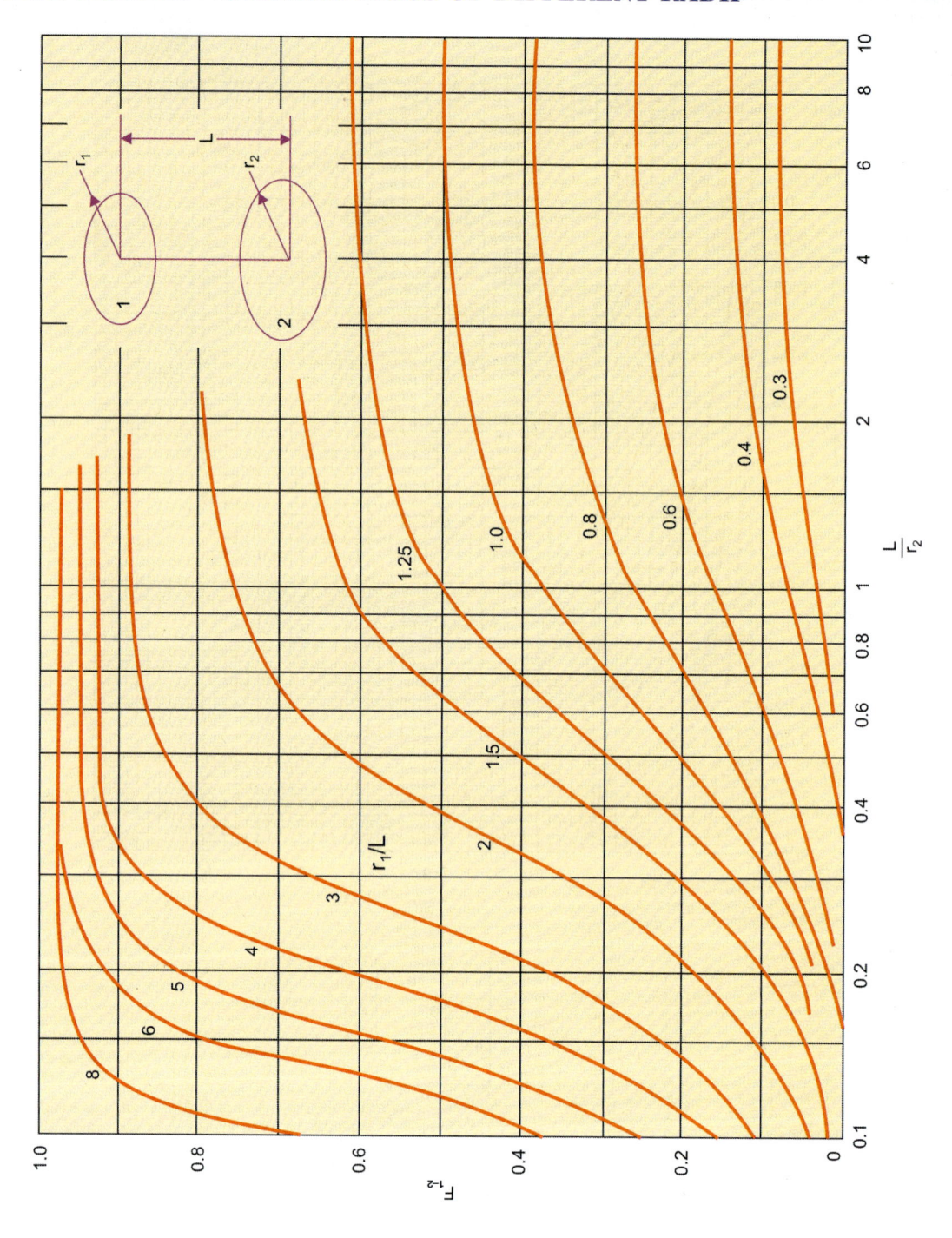

SHAPE FACTOR F_{1-2}—FOR SLOTS AND OPENINGS IN WALL OF THICKNESS, L

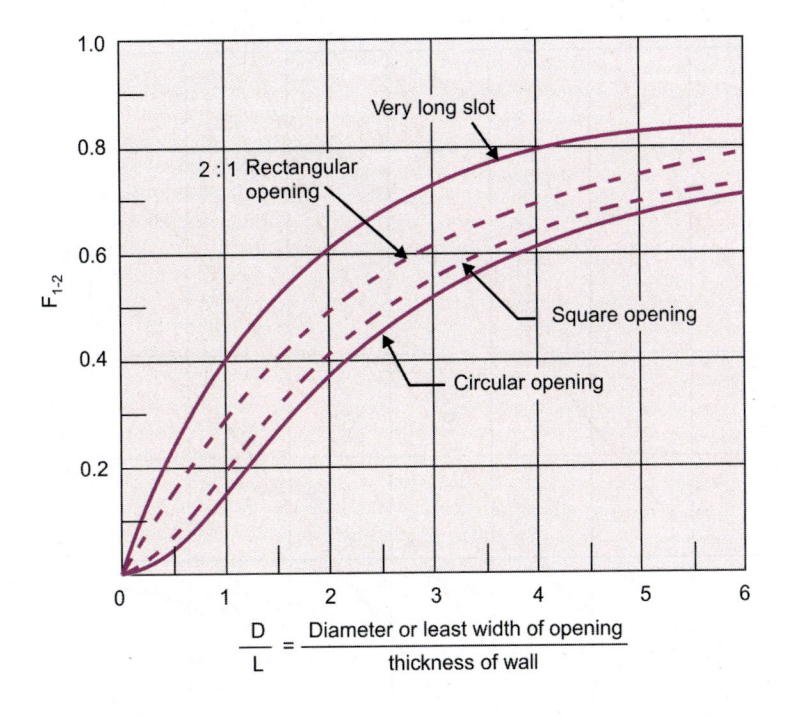

Graph of F_{1-2} (y-axis, 0 to 1.0) versus $\frac{D}{L} = \frac{\text{Diameter or least width of opening}}{\text{thickness of wall}}$ (x-axis, 0 to 6), showing curves for:
- Very long slot
- 2:1 Rectangular opening
- Square opening
- Circular opening

SHAPE FACTOR FOR A SURFACE ELEMENT dA TO A RECTANGULAR SURFACE PARALLEL TO IT AT CORNER

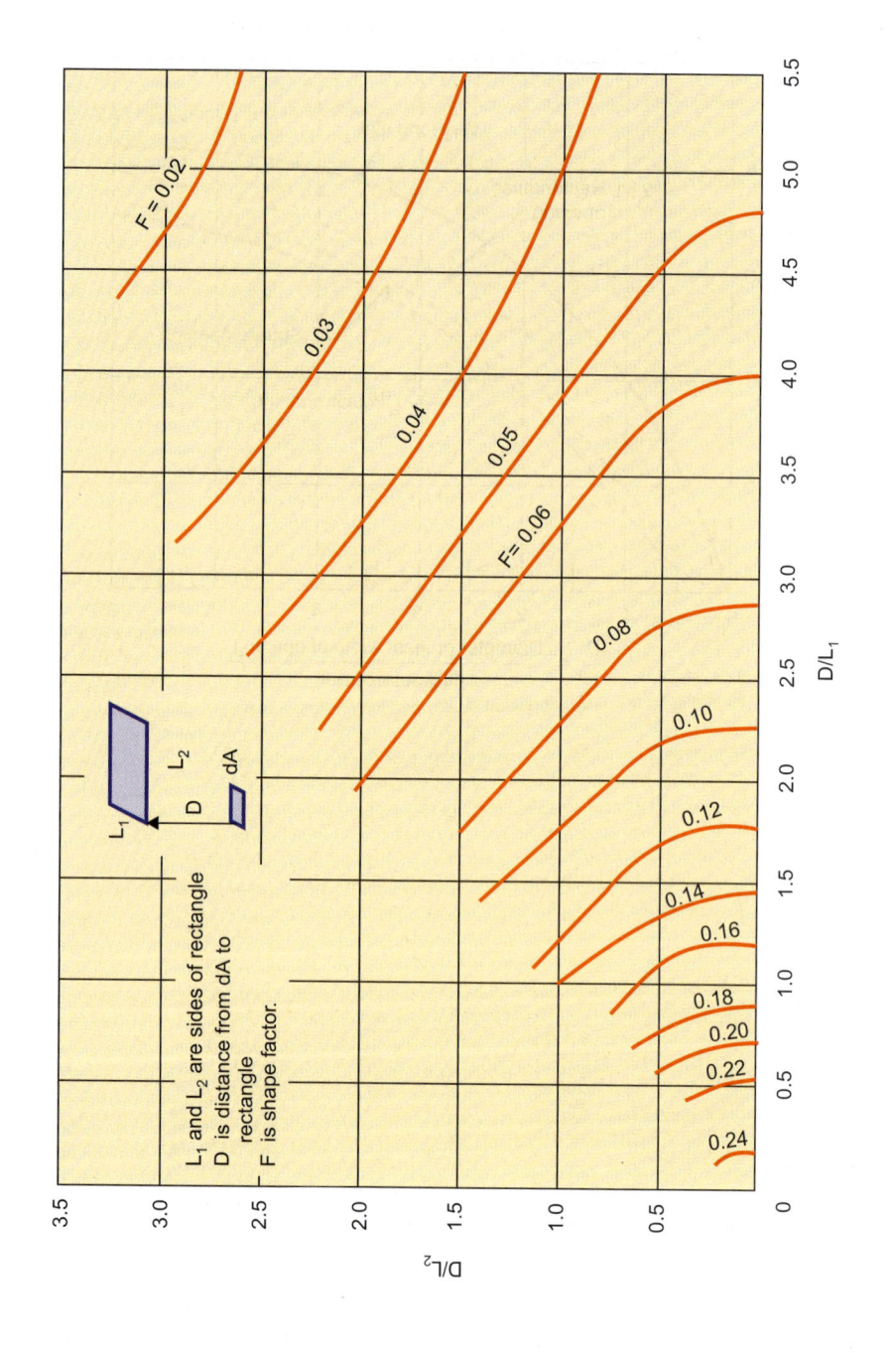

HEAT AND MASS TRANSFER DATA BOOK

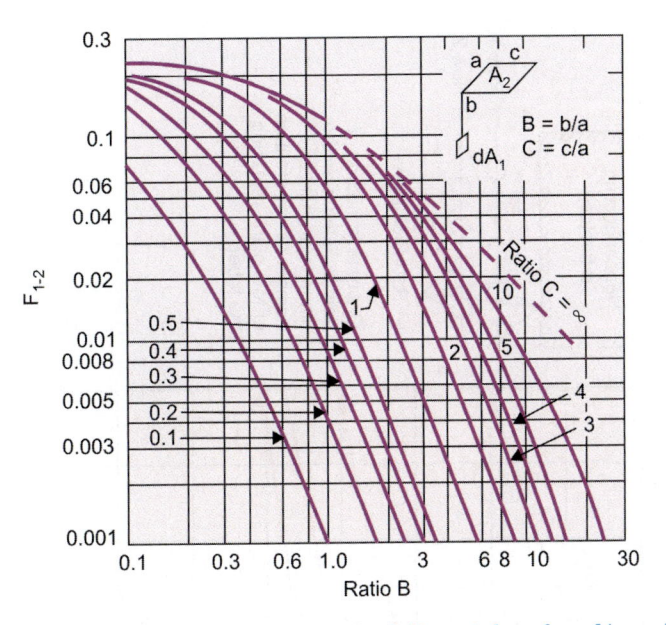

Shape Factor for a system of a differential surface dA_1 and a finite rectangular surface perpendicular to it

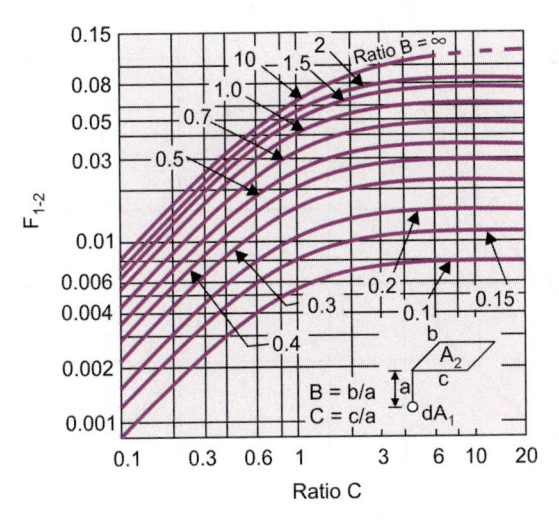

Shape Factor for a system of a differential spherical surface dA_1 and a finite rectangular surface

SHAPE FACTOR COMPARED TO PARALLEL PLANES, FOR A PLANE AND ONE OR TWO ROWS OF TUBES PARALLEL TO IT

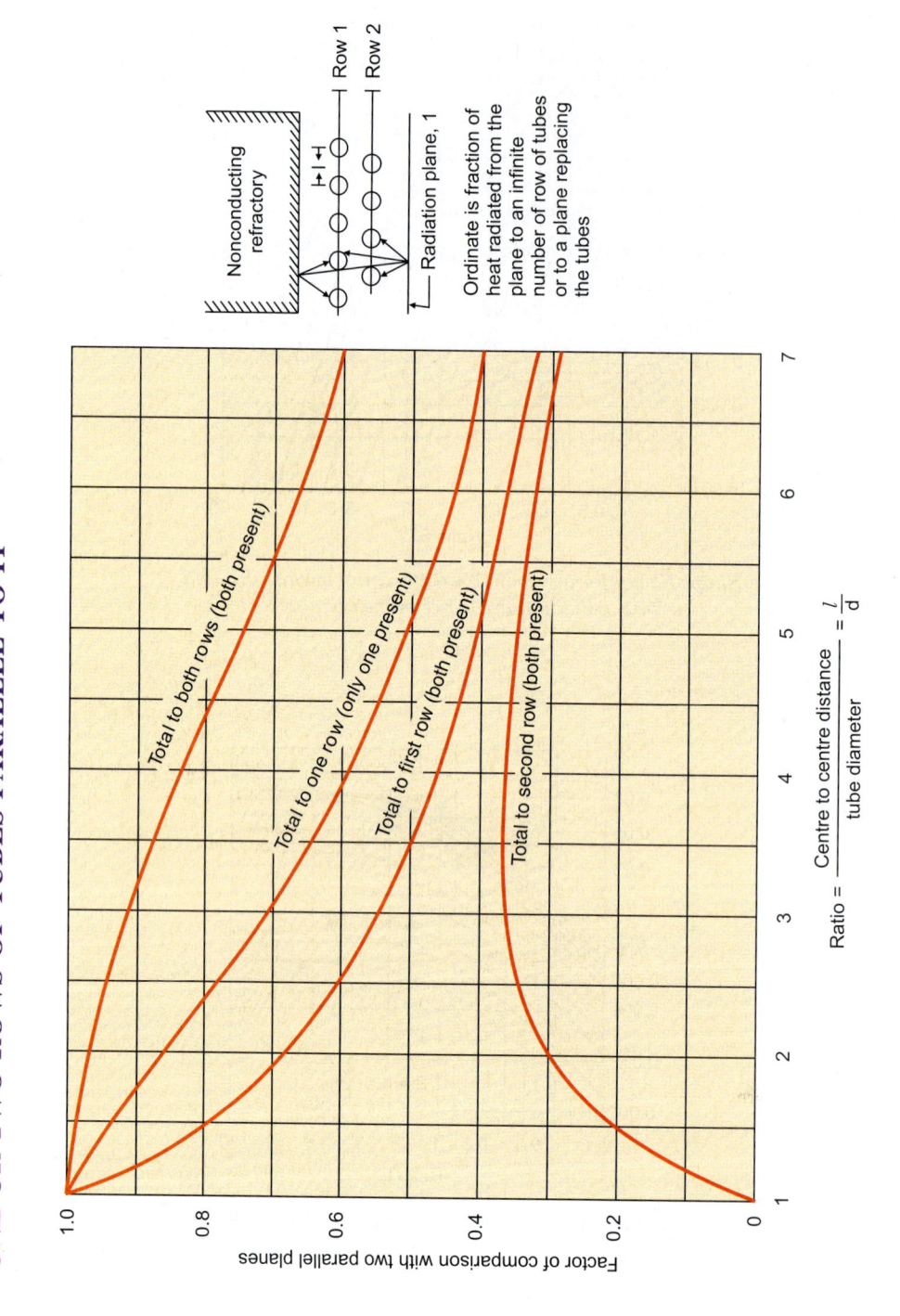

SHAPE FACTOR FOR RADIATION TO BANKS OF TUBES IN VARIOUS ARRANGEMENTS

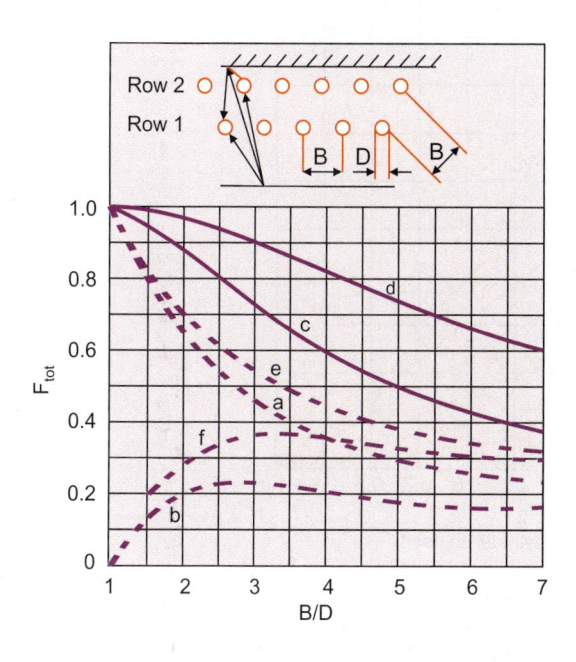

Shape Factor for radiation to a bank of tubes in staggered arrangement— **Equilateral Triangle**

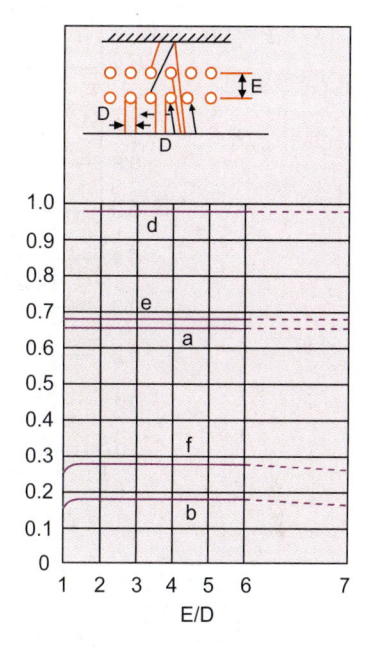

Shape Factor for radiation to a bank of tubes arranged in line

Curve a : Direct radiation to row 1
Curve b : Direct radiation to row 2
Curve c : Total radiation to row 1 if bank consists of one row only
Curve d : Total radiation to rows 1 and 2 ⎫
Curve e : Total radiation to row 1   ⎬ if bank consists of two rows
Curve f : Total radiation to row 2   ⎭

SHAPE FACTOR—SHORT CONCENTRIC CYLINDERS

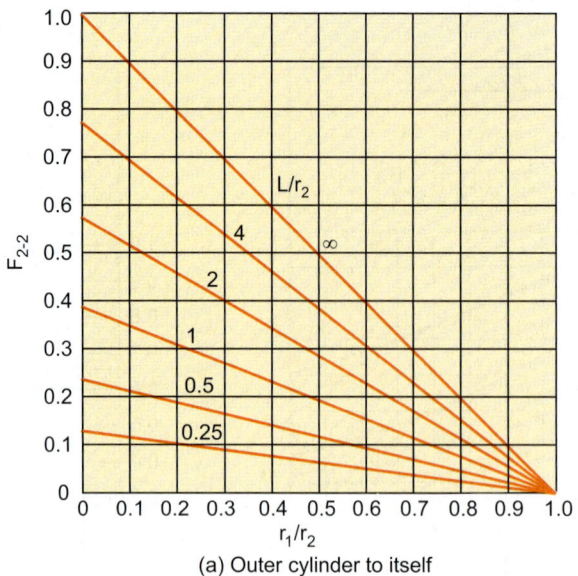

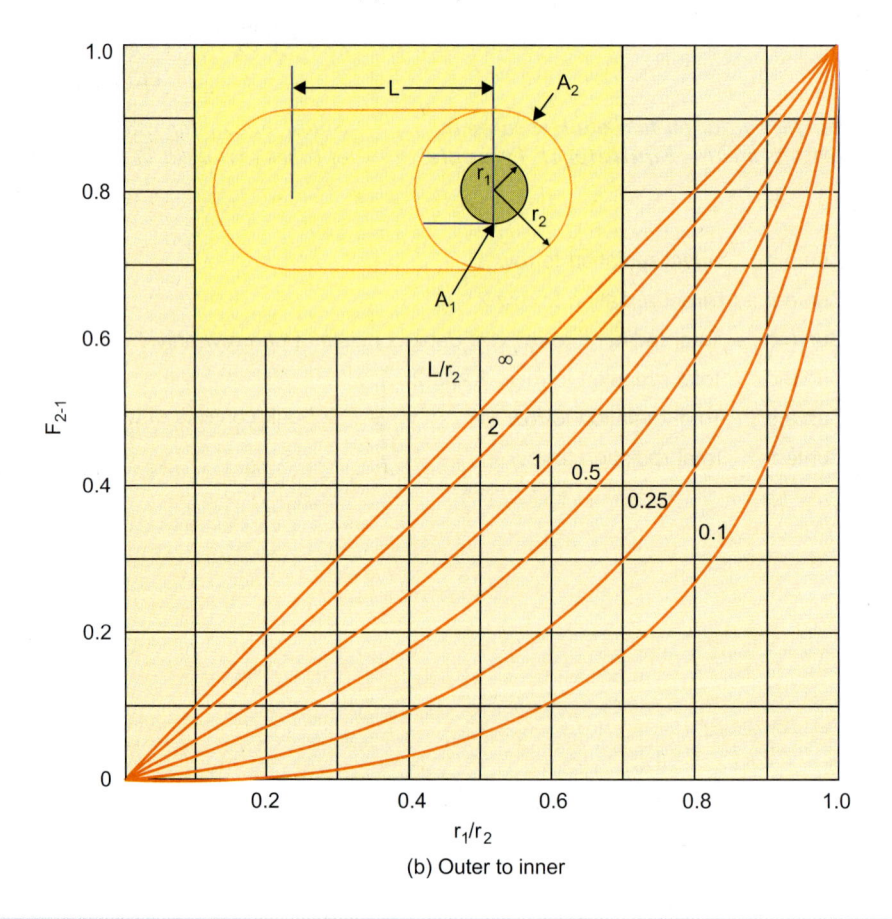

(a) Outer cylinder to itself

(b) Outer to inner

GAS RADIATION

Condition of Gas or Mixture of Gases	*Emissivity, ε_R
1. Carbon dioxide in mixture (i) Only CO_2 as radiating gas in the mixture at 1 atm. (ii) Mixture pressures other than 1 atm.	Refer chart on page 106 for ε_{CO_2} Refer chart on page 107 for correction factor, C_{CO_2}, by which ε_{CO_2} has to be multiplied
2. Water Vapour in mixture (i) at a total pressure of one atm. and near zero partial pressure. (ii) Mixture pressures other than one atm.	Refer page 108 for ε_{H_2O} Refer chart on page 109 for correction factor C_{H_2O} by which ε_{H_2O} has to be multiplied
3. Correction $\Delta\varepsilon$ for mixture when both water vapour and carbon dioxide are present in the mixture	Refer to chart on page 110 $\varepsilon_g = \varepsilon_{H_2O} + \varepsilon_{CO_2} - \Delta\varepsilon$ ε_{H_2O} and ε_{CO_2} have to be determined for the particular cases.

$\dfrac{Q}{A} = \sigma[\varepsilon_s T_s^4 - \alpha_g T_w^4]$ for radiant heat transfer between gas at T_g and enclosure at T_w.

Determine α_g at T_w and ε_g at T_g,

$\alpha_g = \varepsilon_{CO_2} \left(\dfrac{T_g}{T_w}\right)^{0.65} + \varepsilon_{H_2O} \left(\dfrac{T_g}{T_w}\right)^{0.45} - \Delta\varepsilon$, To find α_g use $PL\dfrac{T_w}{T_g}$ instead of PL

Determine ε_{CO_2}, ε_{H_2O} and $\Delta\varepsilon$ at T_w, in this case.

*For non-luminous gas radiation layers of different shapes, an equivalent thickness given on page 111 should be used for L in the charts on pp. 106 to 110.

EMISSIVITY OF CARBON DIOXIDE IN A MIXTURE AT 1 atm

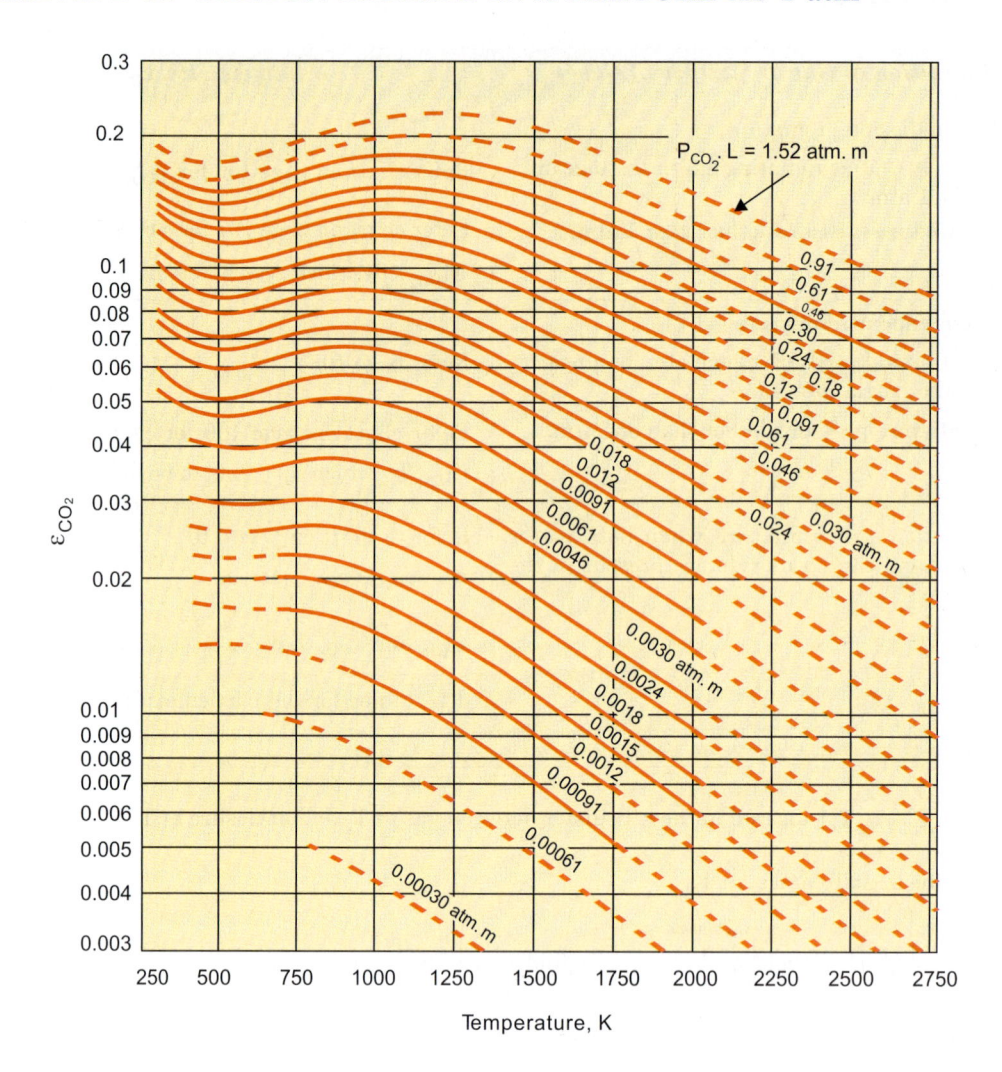

For L refer page 111.

CORRECTION FACTOR FOR C_{CO_2} FOR EMISSIVITY OF CARBON DIOXIDE WHEN MIXTURE PRESSURE IS OTHER THAN 1 atm

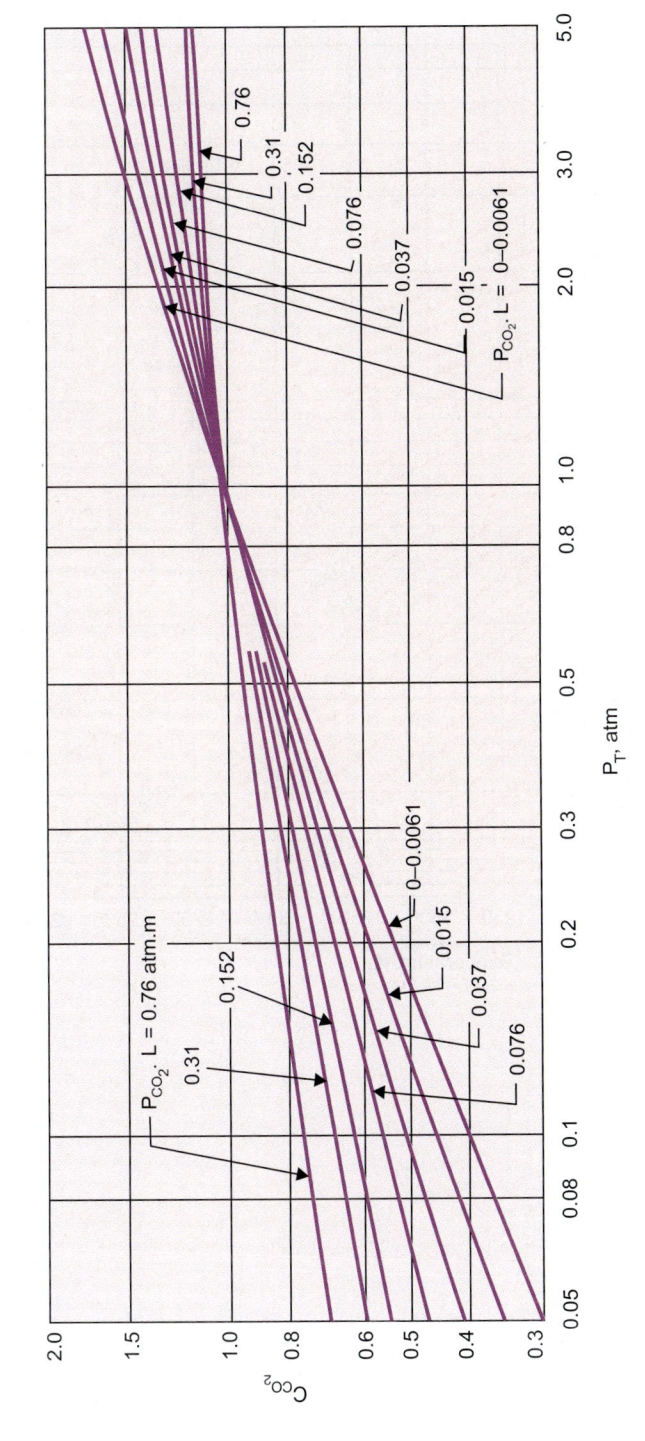

For L refer page 111.

EMISSIVITY OF WATER VAPOUR IN A MIXTURE AT 1 atm

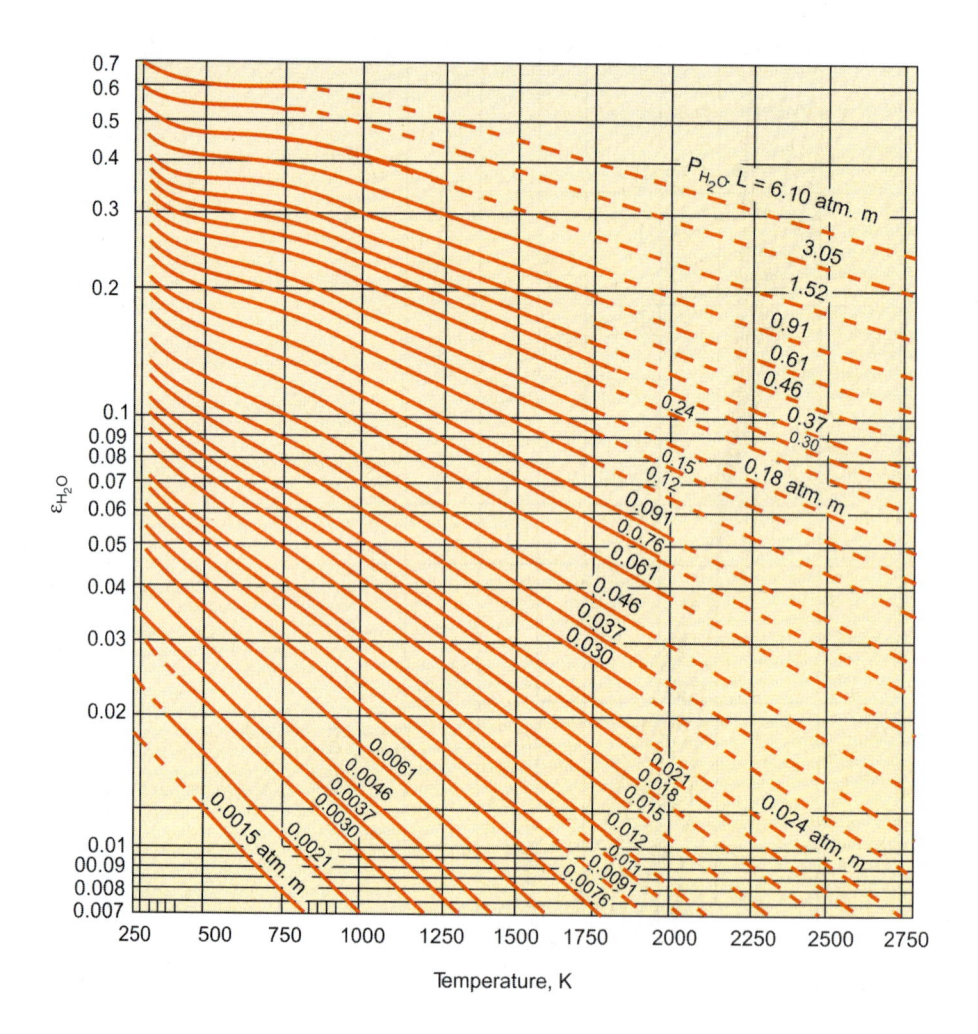

*For L refer page 111.

CORRECTION FACTOR C_{H_2O} FOR ε_{H_2O} WHEN MIXTURE PRESSURE IS OTHER THAN 1 atm

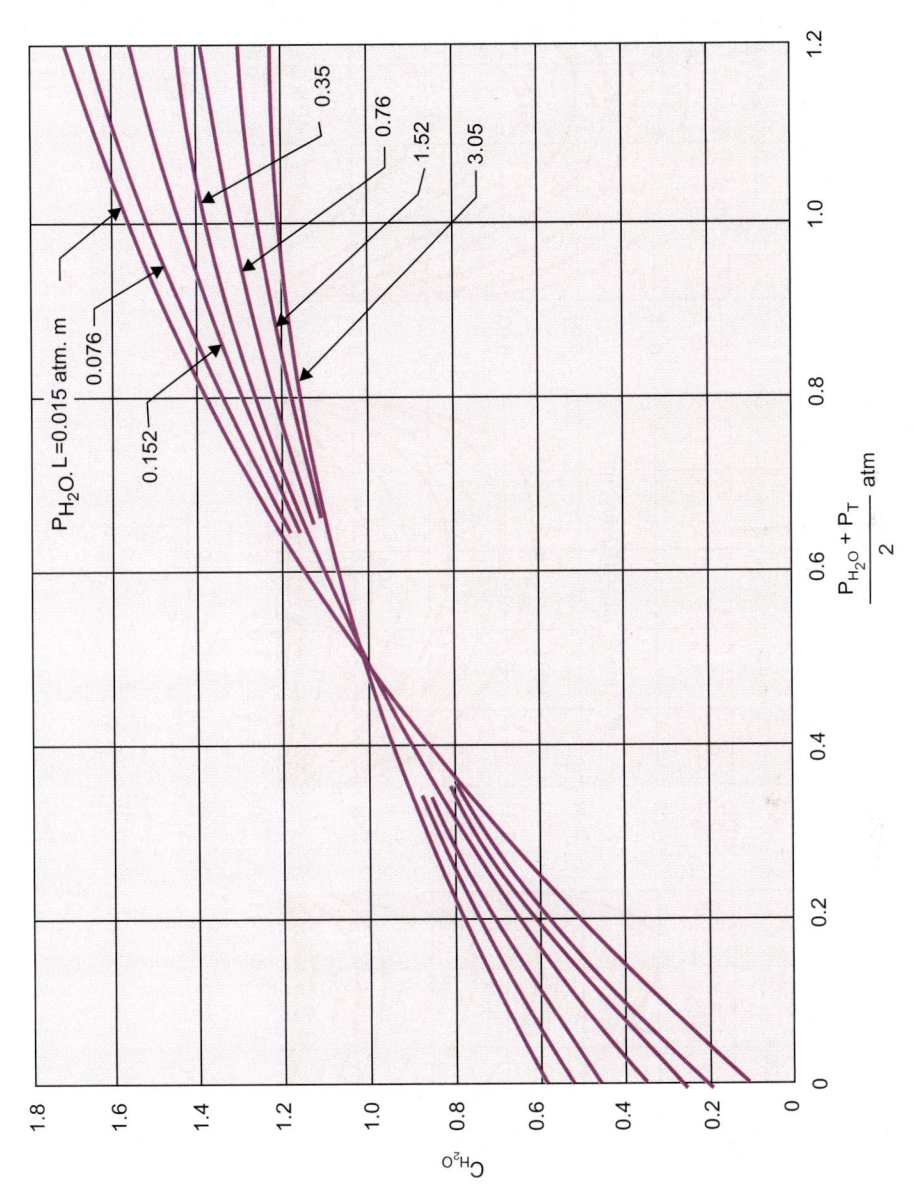

*For L refer page 111.

CORRECTION FACTOR FOR EMISSIVITY WHEN BOTH CO_2 AND H_2O ARE PRESENT IN THE MIXTURE

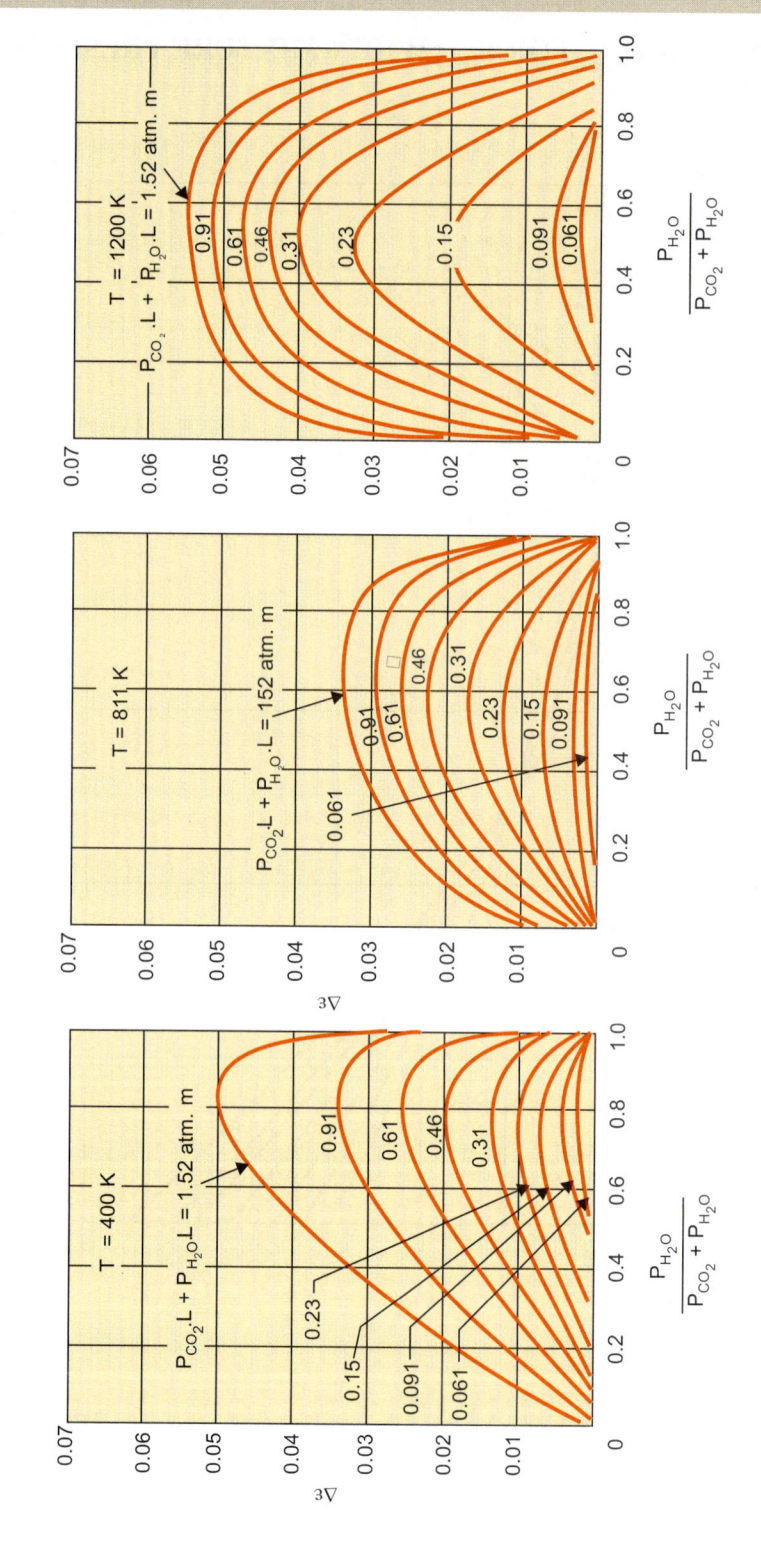

For L refer page 111.

EQUIVALENT THICKNESS, L FOR NON-LUMINOUS GAS RADIATION LAYERS OF DIFFERENT SHAPES

Shape	Characteristic Dimension Z	Factor by which Z is to be Multiplied to Give Equivalent L for Hemispherical Radiation – Calculated by Various Workers	$3.4 \times$ (volume/area)
Sphere	Diameter	0.60	0.57
Cube	Side	0.60	0.57
Infinite cylinder radiating to walls	Diameter	0.90	0.85
Infinite cylinder radiating to centre of base	Diameter	0.90	0.85
Cylinder, height = diameter, radiating to whole surface	Diameter	0.60	0.57
Cylinder, height = diam., radiating to centre of base	Diameter	0.77	0.57
Space between infinite parallel planes	Distance apart	1.80	1.70
Space outside infinite bank of tubes with centres on equilateral triangles, tube diameter = clearance	Clearance	2.80	2.89
Same as above, with tube diameter = one half clearance	Clearance	3.80	3.78
Same as above, with tube centres on squares, and tube diameter = clearance	Clearance	3.50	3.49
Rectangular parallelepiped, $1 \times 2 \times 6$ radiating to :			
2×6 face	Shortest edge	1.06	1.01
1×6 face	Shortest edge	1.06	1.05
1×2 face	Shortest edge	1.06	1.01
all faces	Shortest edge	1.06	1.02
Infinite cylinder of semicircular cross-section radiating to centre of flat side	Diameter	0.63	0.52

DIMENSIONLESS GROUPS : (D—DIFFUSION COEFFICIENT, d—DIAMETER)

Groups	Symbol	Name	Significance
hL/k_s, hR/k	Bi	Biot Number	$\dfrac{\text{Internal conduction resistance}}{\text{Surface convection resistance}}$
$\Delta P/\rho u^2$	Eu	Euler Number	Pressure force / Inertia force
$\dfrac{\alpha\tau}{R^2}$, $\dfrac{\alpha\tau}{L^2}$	Fo	Fourier Number	$\dfrac{\text{Characteristic body dimension}}{\text{Temperature wave penetration depth in time, }\tau}$
u^2/gL	Fr	Froude Number	Inertia force / gravity force
$RePr\dfrac{d}{x}$	Gz	Graetz Number	$\dfrac{\text{Heat transfer by conduction}}{\text{Heat transfer by convection}}$ (with entrance region consideration)
$g\beta\Delta TL^3\rho^2/\mu^2$	Gr	Grashof Number	$\left(\dfrac{\text{Buoyant force}}{\text{Viscous force}}\right)\left(\dfrac{\text{Inertia force}}{\text{viscous force}}\right)$
λ/L	Kn	Knudson Number	$\dfrac{\text{Mean free path}}{\text{Characteristic body dimension}}$
α/D, Sc/Pr	Le	Lewis Number	$\dfrac{\text{Heat diffusivity}}{\text{Mass diffusivity}}$
$\dfrac{hL}{k}$, $\dfrac{hd}{k}$	Nu	Nusselt Number	Ratio of temperature gradients by conduction and convection at the surface
$uL\rho C_p/k$, $uD\rho C_p/k$	$Pe = RePr$	Peclet Number	$\dfrac{\text{Heat transfer by convection}}{\text{Heat transfer by conduction}}$
$\dfrac{C_p\mu}{k}$	Pr	Prandtl Number	$\dfrac{\text{Molecular diffusivity of momentum}}{\text{Molecular diffusivity of heat}}$
$\dfrac{g\beta\Delta TL^3\rho^2 C_p}{\mu k}$	$Ra = GrPr$	Rayleigh Number	Refer Pr and Gr
$\dfrac{uL}{\nu}$, $\dfrac{ud}{\nu}$, $\dfrac{ud\rho}{\mu}$	Re	Reynolds Number	Inertia force / viscous force
$\mu/\rho D$, $\dfrac{\nu}{D}$	Sc	Schmidt Number	$\dfrac{\text{Molecular diffusivity of momentum}}{\text{Molecular diffusivity of mass}}$
$h_m L/D$, $h_m d/D$	Sh	Sherwood Number	Ratio of concentration gradients at the boundary by diffusion and by convection
$\dfrac{h}{c_p\rho u}$	$St = \dfrac{Nu}{RePr}$	Stanton Number	$\dfrac{\text{Wall heat transfer rate}}{\text{Heat transfer by convection}}$

EXTERNAL FLOW

Note: Properties are to be evaluated (unless otherwise started) at film temperature $T_f = (T_w + T_\infty)/2$.

1. Flat Plate

Flow Conditions	Correlation and Validity	Notations
1.0 LAMINAR FLOW $(Re_x < 5 \times 10^5)$ Boundary Layer Thicknesses: Friction Factor:	$\delta_{hx} = 5x\, Re_x^{-0.5}$ $\delta_{Tx} = \delta_{hx}\, Pr^{-0.333}$ $\delta_x = \delta_{hx}/3$ $\delta_{ix} = \delta_{hx}/7$ $C_{fx} = 0.664\, Re_x^{-0.5}$ $\overline{C_{fL}} = 1.328\, Re_L^{-0.5}$	T_f = Film temperature $(T_w + T_\infty)/2$ T_∞ = Free stream fluid temperature T_w = Plate surface temperature Re_x = Reynolds number at a distance x from leading edge. Re_L = Reynolds number at location L δ_{hx} = Hydrodynamic boundary layer thickness at a distance x from leading edge.
1.1 Constant Wall Temperature	$Nu_x = 0.332\, Re_x^{0.5}\, Pr^{0.333}$ $0.6 < Pr < 50$	δ_{Tx} = Thermal boundary layer thickness at a distance x from leading edge.
1.1.1 If heating starts from a distance x_o from leading edge	$Nu_x = 0.332\, Re_x^{0.5}\, Pr^{0.333}$ $\cdot\, [1-(x_o/x)^{0.75}]^{-0.333}$	δ_x = Displacement thickness at x δ_{ix} = Momentum thickness at x C_{fx} = Local friction coefficient
1.1.2 For liquid metals (low Prandtl numbers) and for silicones (High Prandtl numbers)	Nu_x $= \dfrac{(0.3387\, Re_x^{0.5}\, Pr^{0.333})}{[(1+0.0468/Pr)^{0.67}]^{0.25}}$ $Pr < 0.05$, $Re_x\, Pr > 100$, also for $Pr > 50$	defined by $\tau_{wx}/(\rho u_\infty^2/2)$ (Fanning friction factor) τ_{wx} = Wall shear stress at x N/m^2 $\overline{C}_{fL}$ = Average friction coefficient upto the distance L from leading edge.
1.1.3 For Liquid Metals only	$Nu_x = 0.565\,[Re_x \cdot Pr]^{0.5}$ for $Pr \le 0.05$	Pr = Prandtl number x_o = Distance from leading edge at which heating starts ρ = density
1.2 Constant heat flux	$Nu_x = 0.453\, Re_x^{0.5}\, Pr^{0.333}$, $0.6 < Pr < 50$ $\overline{T_w - T_\infty}$ $= \dfrac{(qL/k)}{[0.6795\, Re_L^{0.5}\, Pr^{0.333}]}$	u = free stream velocity q = heat flux W/m^2 k = Thermal conductivity of fluid. $\overline{T_w - T_\infty}$ = Average temperature difference between plate and fluid.

(Contd.)

Flow Conditions	Correlation and Validity	Notations
1.2.1 for liquid metals or silicones Constant heat flux	$Nu_x = \dfrac{[0.4637\, Re_x^{0.5}\, Pr^{0.333}]}{[1 + (0.0207/Pr)^{0.67}]^{0.25}}$ Valid for : $Pr > 50$ and $Pr \leq 0.05$ and for $Re_x\, Pr > 100$	$\overline{Nu_L}$ = Average Nusselt Number upto length L Nu_L = Nusselt Number at location L
1.3 Average values for constant heat flux or constant wall temperature	$\overline{Nu_L} = 2Nu_L$ $St \cdot Pr^{2/3} = C_f/2$	St = Stanton Number $= \dfrac{Nu}{Re \cdot Pr}$
1.4 **FLAT PLATE TURBULENT FLOW** $5 \times 10^5 > Re_x < 10^7$	$\delta_{hx} = 0.381x\, Re^{-0.2}$ $\delta_{tx} \approx \delta_{hx}$	C_f = Average friction coefficient x = distance from leading edge
1.4.1 Fully Turbulent from leading edge	$\delta_x = \delta_{hx}/8$ $\delta_{ix} = (7/72)\, \delta_{hx}$, $Nu_x = 0.0296\, Re_x^{0.8}\, Pr^{0.33}$ $\overline{Nu_L} = 0.037\, Re^{0.8}\, Pr^{0.33}$ Or by analogy Nu_x $= \dfrac{(C_{fx}/2)\, Re_x\, Pr}{1 + 12.8\,(C_{fx}/2)^{0.5}\,(Pr^{0.68} - 1)}$ $C_{fx} = 0.0592\, Re_x^{-0.2}$: $\quad 5 \times 10^5 < Re < 10^7$ $C_{fx} = 0.37\, [\log_{10} Re_x]^{-2.584}$; $\quad 10^7 < Re < 10^9$	δ_{hx} = hydrodynamic boundary layer thickness at x δ_{tx} = thermal boundary layer thickness at x δ_x = displacement thickness at x δ_{ix} = momentum thickness at x
1.4.2 FLAT PLATE, Turbulent Constant heat flux	$Nu_x = 1.04$ times Nu_x calculated as per constant wall temperature	
1.5 Laminar-Turbulent (combined)	$\delta_{hx} = 0.381x\, Re_x^{-0.2} - 10256x\, Re_x^{-1.0}$	C_{fx} = local friction coefficient defined by $\tau_{wx}/(\rho u_\infty^2/2)$

HEAT AND MASS TRANSFER DATA BOOK

Flow Conditions	Correlation and Validity	Notations
1.5.1 Laminar-Turbulent (Laminar length considered)	$C_{fL} = 0.074\, Re_L^{-0.2} - 1742\, Re_L^{-1.0}$ (Critical Reynolds Number $= 5 \times 10^5$) OR $$C_{fL} = \frac{0.455}{(\log_{10} Re_L)^{2.584}} - \frac{A}{Re_L}$$ where <table><tr><th>Critical Reynolds Number</th><th>Value of A</th></tr><tr><td>3×10^5</td><td>1050</td></tr><tr><td>5×10^5</td><td>1700</td></tr><tr><td>1×10^6</td><td>3300</td></tr><tr><td>3×10^6</td><td>8700</td></tr></table>	C_{fL} = Average friction coefficient upto length L.
1.5.2 Laminar-turbulent constant wall temperature	$\overline{Nu_L} = Pr^{0.333}\,[0.037\, Re_L^{0.8} - 871]$ Critical Reynolds Number $= 5 \times 10^5$ $Re_L \leq 10^8$, $0.6 < Pr < 60$ OR generally $\overline{Nu_L} = Pr^{0.333}\,[0.037\, Re_L^{0.8} - A]$ $A = 0.037\, Re_{cr}^{0.8} - 0.664\, Re_{cr}^{0.5}$	Re_{cr} = Critical Reynolds Number $\overline{Nu_L}$ —Average Nusselt Number upto length L
1.5.3 By Analogy	$St_x \cdot Pr^{2/3} = \dfrac{C_{fx}}{2}$ Properties to be evaluated at T_∞	T_∞ = Free stream temperature
1.6 High speed flow, constant wall temperature	The heat transfer equation to be used is: $q = hA\,[T_w - T_{ad}]$ and h is calculated as in low speed flow, using the properties at a temperature T defined by $T = T_\infty + 0.5(T_w - T_\infty)$ $\qquad + 0.22\,(T_{ad} - T_\infty)$ where T_{ad} is calculated using the recovery factor R $R = \dfrac{T_{ad} - T_\infty}{T_o - T_\infty}$ where $T_o = T_\infty \left[1 + \left(\dfrac{\gamma - 1}{2}\right) M_\infty^2\right]$	γ = ratio of specific heats R = to be specified < 1 M_∞ = free stream Mach Number

(Contd.)

2. FLOW OVER CYLINDERS

Flow Condition	Correlation and Validity	Notations
2.1.1 Across cylinders Generalised equation	@ $Nu_D = C \cdot Re_D^m \cdot Pr^{0.333}$ (for C, m refer below) Properties at $T_f = (T_\infty + T_w)/2$	Nu_D = Nusselt Number based on diameter Re_D = Reynolds Number based on diameter D = Diameter m = Coefficient C = Constant
	Re_D | C | m 0.4–4.0 | 0.989 | 0.330 4.0–40.0 | 0.911 | 0.385 40.0–4000 | 0.683 | 0.466 4000.0–40,000 | 0.193 | 0.618 40,000.0–400,000 | 0.0266 | 0.805	
2.1.2 Flow over non circular shapes	$Nu = 1.1\, C_1\, Re_D^n\, Pr^{0.33}$, $5 \times 10^3 < Re_D < 10^5$ Refer page 119 for C_1 and n	@ Applicable for mass transfer also

FLOW OVER CYLINDERS *(Contd.)*

Flow Condition	Correlation and Validity	Notations
2.1.3 ACROSS CYLINDERS For liquids	$Nu_D = [0.35 + 0.56 \cdot Re_D^{0.52}] Pr^{0.333}$ $10^{-1} < Re_D < 10^5$, Properties at T_f	
For liquids and gases	$Nu = [0.43 + 0.50 Re^{0.5}] Pr^{0.38}$ $\left[\dfrac{Pr_f}{Pr_w}\right]^{0.25}$ $1 < Re < 10^3$	
2.1.4 When property variation is large due to temperature variation	$Nu = 0.25 Re^{0.6} Pr^{0.38} \left[\dfrac{Pr_f}{Pr_w}\right]^{0.25}$ $10^3 < Re < 2 \times 10^5$ gases – properties at T_f (film temp.) liquids – properties at free stream temperature T_∞ $Nu = [0.4 Re_D^{0.5} + 0.06 Re_D^{0.67}]$ $\times Pr^{0.4} \left[\dfrac{\mu_\infty}{\mu_w}\right]^{0.25}$ $10 < Re < 10^5$; $0.67 < Pr < 300$ $0.25 < \left(\dfrac{\mu_\infty}{\mu_w}\right) < 5.3$, Properties for Re, Pr at T_∞	Pr_f = Prandtl Number at film temperature Pr_w = Prandtl Number at wall temperature μ_∞ = Dynamic viscosity at free stream temperature μ_w = Dynamic viscosity at wall temperature
2.1.5 Generalised form	$Nu = C \cdot Re^m Pr^n \cdot \left(\dfrac{Pr_\infty}{Pr_w}\right)^{0.25}$ $0.7 < Pr < 500$ $1 < Re < 10^6$ Properties at T_∞ $n = 0.37$ for $Pr < 10$ $n = 0.36$ for $Pr > 10$	C, m tables (not suitable for liquid metal) | Re | C | m | |---|---|---| | 1 – 40 | 0.75 | 0.4 | | 40 – 1000 | 0.51 | 0.5 | | $10^3 - 2 \times 10^5$ | 0.26 | 0.6 | | $2 \times 10^5 - 10^6$ | 0.076 | 0.7 |

Flow Condition	Correlation and Validity	Notations
2.1.6 Liquid metals and silicones	$$Nu = 0.3 + \frac{0.62\, Re_D^{0.5}\, Pr^{0.333}}{\left[1 + \left(\frac{0.4}{Pr}\right)^{0.67}\right]^{0.25}}$$ $$\times \left[1 + \left(\frac{Re_D}{282000}\right)^{0.625}\right]^{0.8}$$ $10^2 < Re_D < 10^4$: $Pe > 0.2$, wider range of Pr, all fluid properties at T_f $$Nu = 0.3 + \frac{0.62\, Re^{0.5}\, Pr^{0.333}}{\left[1 + \left(\frac{0.4}{Pr}\right)^{0.67}\right]^{0.25}}$$ $$\times \left[1 + \left(\frac{Re_D}{282000}\right)\right]^{0.5}$$ $20000 < Re < 400{,}000$; $Pe > 0.2$ Properties at T_f (All fluids)	Pe. Peclet Number = Re. Pr
2.1.7 Liquid Metals only creeping flow For constant wall temp. For constant heat flux	$Nu = [0.8237 - \ln(Pe^{0.5})]^{-1}$ $Pe < 0.2$, Properties at T_f $Nu = 1.25\,(Pe)^{0.413}$ $\quad\quad 1 < Pe < 100$ $Nu = 1.05\,(Pe)^{0.5}$ $Nu = 1.145\,(Pe)^{0.5}$	

HEAT AND MASS TRANSFER DATA BOOK

FLOW OVER NON CIRCULAR SECTIONS

Flow Direction and Profile	Re_D From	Re_D To	n	C_1
→ ◇ (square, corner) D	5,000	100,000	0.588	0.222
→ ◇ (square, corner) D	2,500	7,500	0.624	0.261
→ ⬭ (ellipse horizontal) D	2,500	15,000	0.612	0.224
→ ⬬ (ellipse vertical) D	3,000	15,000	0.804	0.085
→ ⬢ (hexagon) D	5,000	100,000	0.638	0.138
→ ⬢ (hexagon) D	5,000	19,500	0.638	0.144
→ ⬢ (hexagon) D	19,500	100,000	0.782	0.035
→ ▢ (square) D	5,000	100,000	0.675	0.092
→ ▢ (square) D	2,500	8,000	0.699	0.160
→ ▭ (rectangle with plate) D	4,000	15,000	0.731	0.205

3. FLOW OVER SPHERES

Flow Condition	Correlation and Validity	Notations
FLOW OVER/ACROSS SPHERES 3.1 For various ranges of Reynolds Numbers For Gases	$Nu = 0.37 \cdot Re^{0.6}$; $17 < Re < 70{,}000$ Properties at T_f $Nu = 2 + (0.25\,Re + 3 \times 10^{-4}\,Re^{1.6})^{0.5}$ $100 < Re < 3 \times 10^5$, $Pr \approx 0.71$	all Re based on diameter
For Air	$Nu = 430 + 5 \times 10^{-3}\,Re + 0.025 \times 10^{-9}\,Re^2 - 3.1 \times 10^{-17}\,Re^3$ $3 \times 10^5 < Re < 5 \times 10^6$; $Pr \simeq 0.71$	
For Liquids	$Nu\,(Pr)^{-0.3} = 0.97 + 0.68\,Re^{0.5}$; $1 < Re < 2000$ Properties at T_f	
3.2 Large Property variation Oils and Water	$Nu \cdot Pr^{-0.3} \left(\dfrac{\mu_w}{\mu_\infty}\right)^{0.25}$ $= 1.2 + 0.53 \cdot Re^{0.54}$ $1 < Re < 200{,}000$ Properties at T_∞	
3.2 Large Property Variations Gases and Liquids	$Nu = 2 + (0.4\,Re^{0.5} + 0.06\,Re^{0.67})$ $\times Pr^{0.4} \left(\dfrac{\mu_\infty}{\mu_w}\right)^{0.25}$, $3.5 < Re < 8 \times 10^4$; $0.7 < Pr < 380$, Properties at T_∞ $1.0 < \dfrac{\mu_\infty}{\mu_w} < 3.2$	
3.3 Falling Liquid Drops	$Nu = 2 + 0.6\,Re^{0.5}\,Pr^{0.333}\,[25(x/D)^{-0.7}]$ Properties at T_∞	x — falling distance, measured from rest D — Diameter of droplet.
3.4 Liquid Metal	$Nu = 2 + 0.386\,(Re \cdot Pr)^{0.5}$ $3.56 \times 10^4 < Re < 1.525 \times 10^5$	

4. FLOW ACROSS TUBE BANKS

Flow Condition	Correlation and Validity	Notations
4.1 Number of rows of tubes 10 or more	$Nu = C\, Re^n$ Re to be calculated on the basis of max. fluid velocity V_{max} $N \geq 10$; $2000 < Re < 40000$; Properties at T_f n, C from table page 123	V_{max} = calculated from free stream velocity and the actual flow area allowing for tube obstruction S_t = Pitch transverse to flow S_l = Pitch along the flow N = Number of tubes u_∞ = free stream velocity just before entry
4.2 For less than 10 rows	Multiply the above Nu by C_1 C_1 – see tables page 123 Check : $(S_t - D) < \left\{ 2\sqrt{\left(\dfrac{S_t}{2}\right)^2 + S_l^{\,2}} - 2D \right\}$ In line : $V_{max} = [S_t/(S_t - D)] \cdot u_\infty$ Staggered : maximum of the above and $V_{max} = \dfrac{S_l}{2(S_D - D)} u_\infty$ where, $S_D = \left[S_l^{\,2} + \left(\dfrac{S_t}{2}\right)^2 \right]^{0.5}$	
4.3 Across Tube Banks with 20 or more rows Valid from 4 rows for air	$Nu = C\, Re^m\, Pr^{0.36} \left[\dfrac{Pr_\infty}{Pr_w}\right]^{0.25}$ $N \geq 20$: $0.7 < Pr < 500$; $1000 < Re < 2 \times 10^6$ Properties at T_∞ ; Re based on V_{max}	Pr_∞ = Prandtl at free stream temperature Pr_w = Prandtl at wall temperature.

Condition	Re range	C	m
In line	$10^3 < Re < 2 \times 10^5$	0.27	0.63
In line	$2 \times 10^5 < Re < 2 \times 10^6$	0.021	0.84
Staggered			
$S_t/S_l < 2$	$10^3 < Re < 2 \times 10^5$	$0.35\left(\dfrac{S_t}{S_l}\right)^{0.2}$	0.6
$S_t/S_l > 2$	$10^3 < Re < 2 \times 10^5$	0.4	0.6
	$2 \times 10^5 < Re < 10^6$	0.022	0.84

(Contd.)

Flow Condition	Correlation and Validity	Notations
4.3.1 Liquid Metals	$Nu_D = 4.03 + 0.228 (Pe)^{0.67}$ $20{,}000 < Re < 80{,}000$	$Pe = Re\, Pr$
4.4 Pressure drop and friction factor	$\Delta P = \dfrac{f\, G^2_{max}\, N}{2.09\, \rho g_o} \left[\dfrac{\mu_w}{\mu_b}\right]^{0.14}$ For in line $f = \left[0.044 + \dfrac{0.08\left(\dfrac{S_l}{d}\right)}{\left(\dfrac{S_l - d}{d}\right)^{0.43 + 1.13(d/S_l)}}\right] Re_{max}^{-0.1}$ For staggered $f = \left[0.25 + \dfrac{0.118}{\left(\dfrac{S_l - d}{d}\right)^{1.08}}\right] Re_{max}^{-0.16}$	G_{max} = mass velocity at minimum flow area, kg/m²s ρ = density at t_∞, kg/m³ N = Number of transverse rows ΔP = Pressure drop in Pascal μ_∞ = absolute viscosity at bulk mean temp. μ_w = absolute viscosity at wall temp. d = Tube diameter

HEAT AND MASS TRANSFER DATA BOOK

FLOW ACROSS BANKS OF TUBES

Note: $Nu = C\, Re^n$
The values of C given in table below are for air. For other fluids multiply these tabulated values of C only, by $1.13\, Pr^{0.33}$

In line

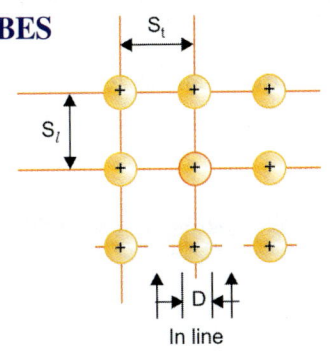

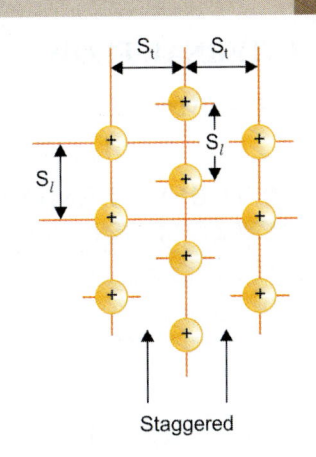

Staggered

FOR TUBE BANKS OF 10 ROWS OR MORE

Arrangement	$\frac{S_t}{D}$	\multicolumn{8}{c	}{S_l/D}
		\multicolumn{2}{c	}{1.25}	\multicolumn{2}{c	}{1.5}	\multicolumn{2}{c	}{2.0}	\multicolumn{2}{c	}{3.0}
		C	n	C	n	C	n	C	n
In Line	1.25	.348	.592	.275	.608	.100	.704	.0633	.752
	1.50	.367	.586	.250	.620	.101	.702	.0678	.744
	2.00	.418	.570	.299	.602	.229	.632	.1980	.648
	3.00	.290	.601	.357	.584	.374	.581	.2860	.608
Staggered	0.6	—	—	—	—	—	—	.213	.636
	0.9	—	—	—	—	.446	.571	.401	.581
	1.0	—	—	.497	.558	—	—	—	—
	1.125	—	—	—	—	.478	.565	.518	.560
	1.25	.518	.556	.505	.554	.519	.556	.522	.562
	1.5	.451	.568	.460	.562	.452	.568	.488	.568
	2.0	.404	.572	.416	.568	.482	.556	.449	.570
	3.0	.310	.592	.356	.580	.440	.562	.421	.574

C_1, Ratio of values of h for N rows deep to that for 10 rows deep $\left[\dfrac{h_N}{h_{10}}\right]$

N	1	2	3	4	5	6	7	8	9	10
Staggered tubes	.68	.75	.83	.89	.92	.95	.97	.98	.99	1.0
In-line tubes	.64	.80	.87	.90	.92	.94	.96	.98	.99	1.0

INTERNAL FLOW

Note : Properties to be evaluated at BULK MEAN temperature

$T_m = (T_{mi} + T_{mo})/2$ unless otherwise stated.

1.0 TUBES : LAMINAR

Flow Condition	Correlation and Validity	Notations
1.1 Tubes, $Re_D < 2300$	$f = \dfrac{64}{Re_D}$, smooth or rough tubes Hydrodynamic entry length $x \simeq 0.04\, D\, Re_D$ Thermal entry length $x \simeq 0.04\, D\, Re_D \cdot Pr$	T_m — Bulk mean temperature $(T_{mi}+T_{mo})/2$ T_{mi} — mean temperature at inlet T_{mo} — mean temperature at outlet $f = \dfrac{\Delta P_f}{\rho \cdot (L/D) \cdot (u_m^2/2)}$; $C_f = f/4$
1.2 Entry Region Hydrodynamic layer fully developed. Thermal layer developing	$Nu = 3.66$ $+ \dfrac{0.0668\,(D/L)\,Re_D\,Pr}{1 + 0.04\,[(D/L)\,Re_D\,Pr]^{0.67}}$ $Pr > 0.7$. Smooth pipe. Constant wall temp. Also refer chart on page 130	ΔP_f — pressure drop due to friction ρ — density L — length D — diameter u_∞ — free stream velocity f — Darcy friction factor
1.2.1 For short lengths for small values of $\dfrac{x}{D}$ constant wall temp.	For $\left(\dfrac{x}{D}\right)/Re_D\,Pr < 0.01$ $Nu = 1.67[Re_D\,Pr/(x/D)]^{0.333}$	x — length u_m — mean velocity in tube
1.2.2 Fully developed Thermal layer	$Nu = 3.66,\ \ L \gg D$ constant wall temp. refer chart on page 115	Re_D — Reynolds Number based on diameter μ — dynamic viscosity of fluid at bulk mean temperature
1.2.3 Simultaneous development of Hydrodynamic and Thermal layers :	$Nu = 3.66$ $+ \dfrac{0.104\,(Re_D\,Pr\cdot D/x)}{1 + 0.16\,(Re_D\,Pr\cdot D/x)^{0.8}}$ $Pr > 0.6$, constant wall temperature	μ_w — dynamic viscosity of fluid at wall temperature $Re_D = \dfrac{4m}{\pi D \mu}$ m — mass flow rate, kg/s
1.2.4 As above for liquids	$Nu = 1.86\left[\dfrac{Re_D\,Pr}{x/D}\right]^{0.333}$ $\times \left(\dfrac{\mu}{\mu_w}\right)^{0.14}$, $RePr\,\dfrac{D}{x} > 10$, $Pr > 0.6$, constant wall temperature	C_f — coefficient of friction

HEAT AND MASS TRANSFER DATA BOOK

Flow Condition	Correlation and Validity	Notations
1.2.5 For non circular tubes, fully devloped flow	Refer page 128 for values of Nusselt numbers	
1.2.6 Short Tubes Thermal developing, hydrodynamic fully developed	$Nu = 1.30 \left[\dfrac{Re_D \, Pr}{x/D}\right]^{0.333}$ $Re \, Pr \, \dfrac{D}{x} > 10$, constant heat flux $Pr > 0.6$. Also see chart on page 130	
1.2.7 Fully developed: constant heat flux	$Nu = 4.36$, $Pr > 0.6$	
1.2.8 Colburn analogy	$St \, Pr^{0.67} = f/8$	

2.0 TUBES : TURBULENT

Flow Condition	Correlation and Validity	Notations
$Re_D > 2300$ 2.1 Friction factors: Smooth tubes	$f = 0.25[1.82 . \log_{10} Re_D - 1.64]^{-2}$ or $f = 0.184 \, Re_D^{-0.2} : Re_D > 10^4$	ε — surface roughness f — friction factor, also refer chart on page 132
2.1.1 Rough tubes	$f = \dfrac{1.325}{\left[\ln\left(\dfrac{\varepsilon}{3.7D} + 5.74/Re^{0.9}\right)\right]^2}$ $10^{-6} \leq \dfrac{\varepsilon}{D} < 10^{-2}$; $5000 < Re < 10^8$	
2.1.2 for transitional turbulent flow	$f = 0.316 . Re_D^{-0.25}$	
2.2.1 Entry length:	$10 \leq \dfrac{x}{D} < 60$	

(Contd.)

Flow Condition	Correlation and Validity	Notations
2.2.2 Entrance region	$Nu = 0.036 \, Re_D^{0.8} \, Pr^{0.33} \times (D/L)^{0.055}$ $10 < \dfrac{L}{D} < 400$ See also figure on page 131	
2.2.3 Short Pipe	$\overline{Nu} = Nu \left(1 + \dfrac{C}{x/D}\right) \, ; \, \dfrac{x}{D} \geq 10$ $C = 1.4$ for hydrodynamic fully developed $C = 6.0$ for hydrodynamic not developed $\overline{Nu} = Nu \, [1 + (D/x)^{0.7}] \, ; \, 2 < \dfrac{x}{D} < 20$	Nu — evaluated at fully developed condition $\overline{Nu}$ — average Nusselt number for the short tube
2.3.1 Fully developed flow Dittus—Boelter eq.	$Nu = 0.023 \, Re_D^{0.8} \cdot Pr^n$ $n = 0.4$ for heating of fluids $n = 0.3$ for cooling of fluids $\quad 0.6 < Pr < 100, \, 2500 < Re < 1.25$ $\quad \times 10^6 \, ; \, \dfrac{L}{D} > 60$ Generally used equation, independent of thermal boundary. Properties at bulk mean temp.	
2.3.2 Smooth tubes : Fully developed Large property variations. Sieder—Tate eq.	$Nu = 0.027 \, Re_D^{0.8} \, Pr^{0.333} \left(\dfrac{\mu_m}{\mu_w}\right)^{0.14}$ $0.7 < Pr < 16700 \, ; \, Re_D \geq 10{,}000 \, ;$ $\dfrac{L}{D} \geq 60$	μ_m — dynamic viscosity at T_m
2.3.3 Fully developed turbulent $1 < \dfrac{\mu_m}{\mu_w} < 40$	$Nu = \dfrac{(f/8) Re_D \, Pr}{1.07 + 12.7(f/8)^{0.5} \, [Pr^{0.67} - 1]}$ $\quad \times \left(\dfrac{\mu_m}{\mu_w}\right)^n$ $n = 0.11$ for heating of fluids $\quad\quad\quad\quad\quad\quad$ constant wall temp. $n = 0.25$ for cooling of fluids $\quad\quad\quad\quad\quad\quad$ constant wall temp. $n = 0$ for constant heat flux or gases Rough/smooth tubes properties at film temp. $10^4 < Re < 5 \times 10^6$ $0.5 < Pr < 200 \, ; \, 6$ per cent accuracy. $200 < Pr < 2000 \, ; \, 10$ per cent accuracy. Replace $\dfrac{\mu_m}{\mu_w}$ by $\dfrac{T_w}{T_m}$ for gases.	μ_w — dynamic viscosity at wall temperature T_w T_w — wall temperature f — calculated by equation given earlier in 2.1 or from chart on page 132

HEAT AND MASS TRANSFER DATA BOOK

Flow Condition	Correlation and Validity	Notations
2.3.4 Rough Tubes: By Colburn analogy	$St_b \cdot Pr_m^{0.67} = \dfrac{f}{8}$	St_b — Stanton Number at bulk temperature Pr_m — Prandtl Number at T_m
2.4.1 Liquid Metals: Constant wall temperature	$Nu = 5 + 0.025\,[Re_D\,Pr]^{0.8}$ $Re_D\,Pr > 100 : \dfrac{L}{D} > 60$	
2.4.2 Constant Heat Flux	$Nu = 4.82 + 0.0185\,(Re_D\,Pr)^{0.827}$ $3.6 \times 10^3 < Re_D < 9.05 \times 10^5$ $10^2 < Re_D\,Pr < 10^4$	
2.4.3 Smooth Tubes, fully developed As above but simpler	$Nu = 0.625\,[Re_D\,Pr]^{0.4}$ $10^2 < Re_D\,Pr < 10^4 : \dfrac{L}{D} > 60$ Constant heat flux	
2.4.4 For Thermal entry region.	$Nu = 3.0\,Re_D^{0.0833}$, $Re_D\,Pr < 100$	
2.5 Non-Circular Sections	The correlation listed above to be used with $D_h = \dfrac{4A}{P}$ where D_h is known as hydraulic diameter.	A — Flow area P — Wetted perimeter

(Contd. on p. 129)

NUSSELT NUMBER FOR FULLY DEVELOPED INTERNAL FLOW, LAMINAR

Geometry ($L/D_h > 100$)		Nu_1 q = constant heat flux	Nu_2 T_w = const.
(circle)		4.364	3.657
(ellipse, 2b vertical, 2a horizontal)	$\dfrac{2b}{2a} = 0.9$	5.331	4.439
(square)	$\dfrac{2b}{2a} = 1$	3.608	2.976
(rectangle)	$\dfrac{2b}{2a} = \dfrac{1}{2}$	4.123	3.391
(rectangle)	$\dfrac{2b}{2a} = \dfrac{1}{4}$	5.099	3.66
(rectangle)	$\dfrac{2b}{2a} = \dfrac{1}{8}$	6.490	5.597
(triangle, 60°)	$\dfrac{2b}{2a} = \dfrac{\sqrt{3}}{2}$	3.111	2.47
(hexagon)		4.002	3.34

HEAT AND MASS TRANSFER DATA BOOK

Flow Condition	Correlation and Validity	Notations
2.6 Concentric tube annulus	$D_h = D_o - D_i$ $Nu_i = \dfrac{h_i D_h}{k}$; $Nu_o = \dfrac{h_o D_h}{k}$	D_o = outside tube diameter D_i = Inside tube diameter h_i = convective heat transfer coefficient on inside surface h_o = convective heat transfer coefficient on outside surface q_i = heat flux on inner surface q_o = heat flux on outer surface Nu_i = Nusselt Number on inside surface, outside adiabatic Nu_o = Nusselt Number on outside surface, inside adiabatic D_h = hydraulic mean diameter $= \dfrac{4 \times \text{flow area}}{\text{Wetted Perimeter}, p}$ $p = \pi(D_o + D_i)$
2.6.1 Fully developed Laminar Constant wall temp.	Values are tabulated below: | D_i/D_o | 0 | 0.05 | 0.10 | 0.25 | 0.50 | 1.0 | |---|---|---|---|---|---|---| | Nu_i | — | 17.46 | 11.56 | 7.37 | 5.74 | — | | Nu_o | 3.66 | 4.06 | 4.11 | 4.23 | 4.43 | 4.86 |	
2.6.2 Constant heat flux Flux can be negative or positive depending on the flow.	$Nu_i = \dfrac{A_i}{1-(q_o/q_i)B_i}$; $Nu_o = \dfrac{A_o}{1-(q_i/q_o)B_o}$ Values are tabulated below: | D_i/D_o | 0 | 0.05 | 0.10 | 0.20 | 0.40 | 0.60 | 0.80 | 1.0 | |---|---|---|---|---|---|---|---|---| | A_i | — | 17.81 | 11.91 | 8.50 | 6.58 | 5.91 | 5.58 | 5.39 | | B_i | ∞ | 2.18 | 1.38 | 0.91 | 0.60 | 0.47 | 0.40 | 0.35 | | A_o | 4.36 | 4.79 | 4.83 | 4.83 | 4.98 | 5.10 | 5.24 | 5.39 | | B_o | 0 | 0.029 | 0.056 | 0.104 | 0.182 | 0.246 | 0.299 | 0.346 |	
2.6.3 Turbulent	Same as for **flow inside tubes** except D_h is to be used in place of 'D'	

LAMINAR ENTRY REGION NUSSELT NUMBERS

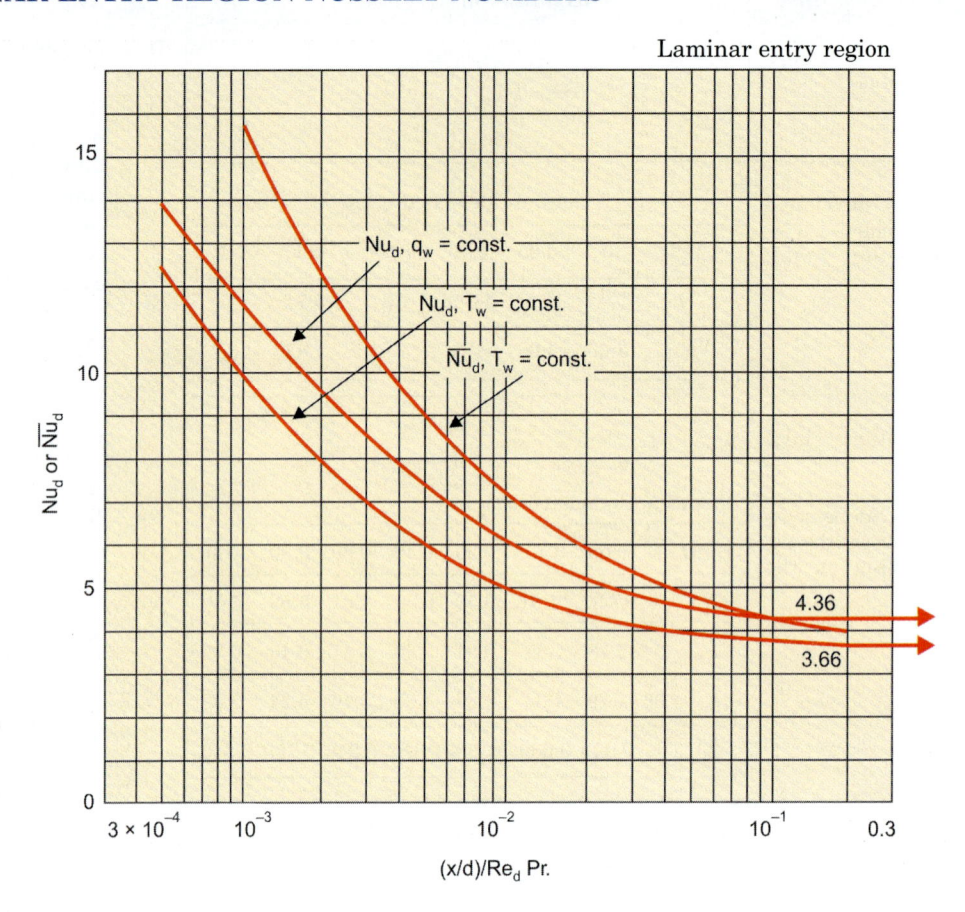

Laminar entry region

Curves shown:
- Nu_d, q_w = const.
- Nu_d, T_w = const.
- $\overline{Nu}_d$, T_w = const.

Asymptotic values: 4.36 and 3.66

Axes: Nu_d or $\overline{Nu}_d$ versus $(x/d)/Re_d\, Pr$

TURBULENT FLOW ENTRY REGION NUSSELT NUMBERS

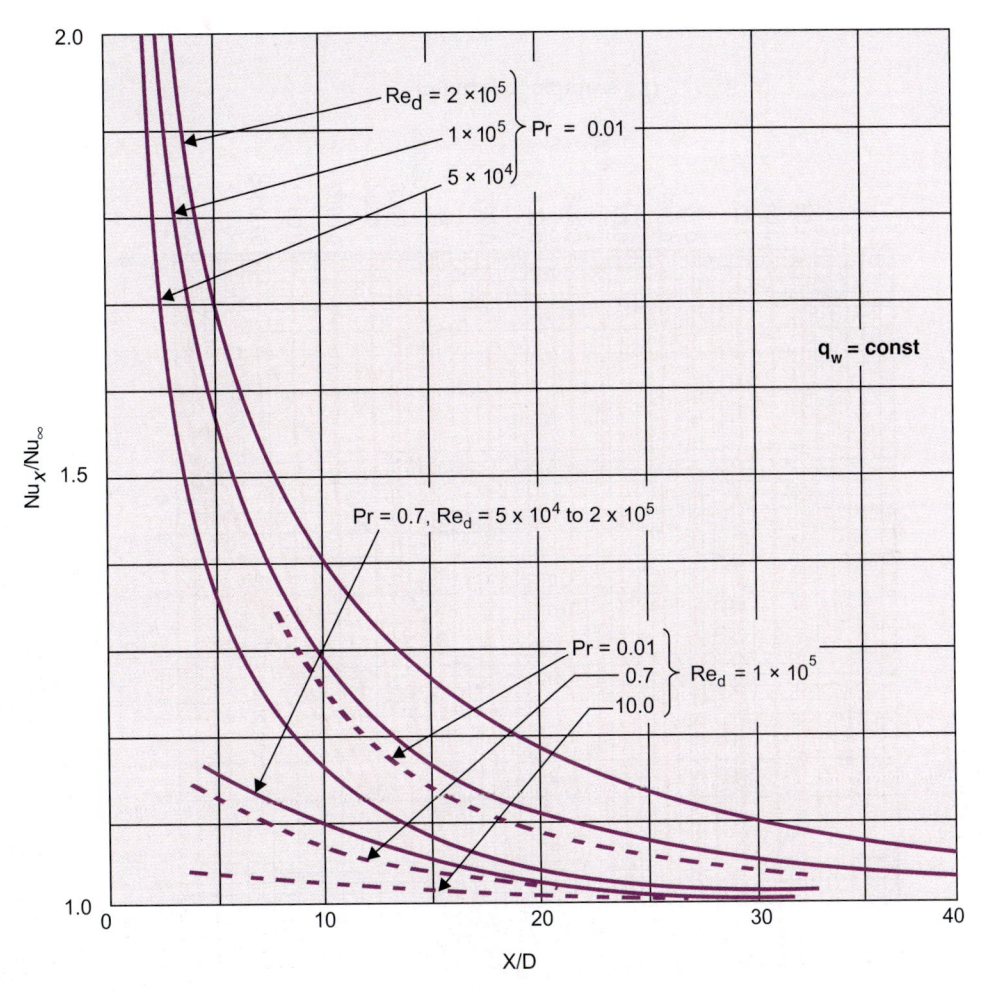

Nu_x — Local Nusselt Number
Nu_∞ — Nusselt Number at fully developed region

FRICTION FACTOR *(See p. 125)*

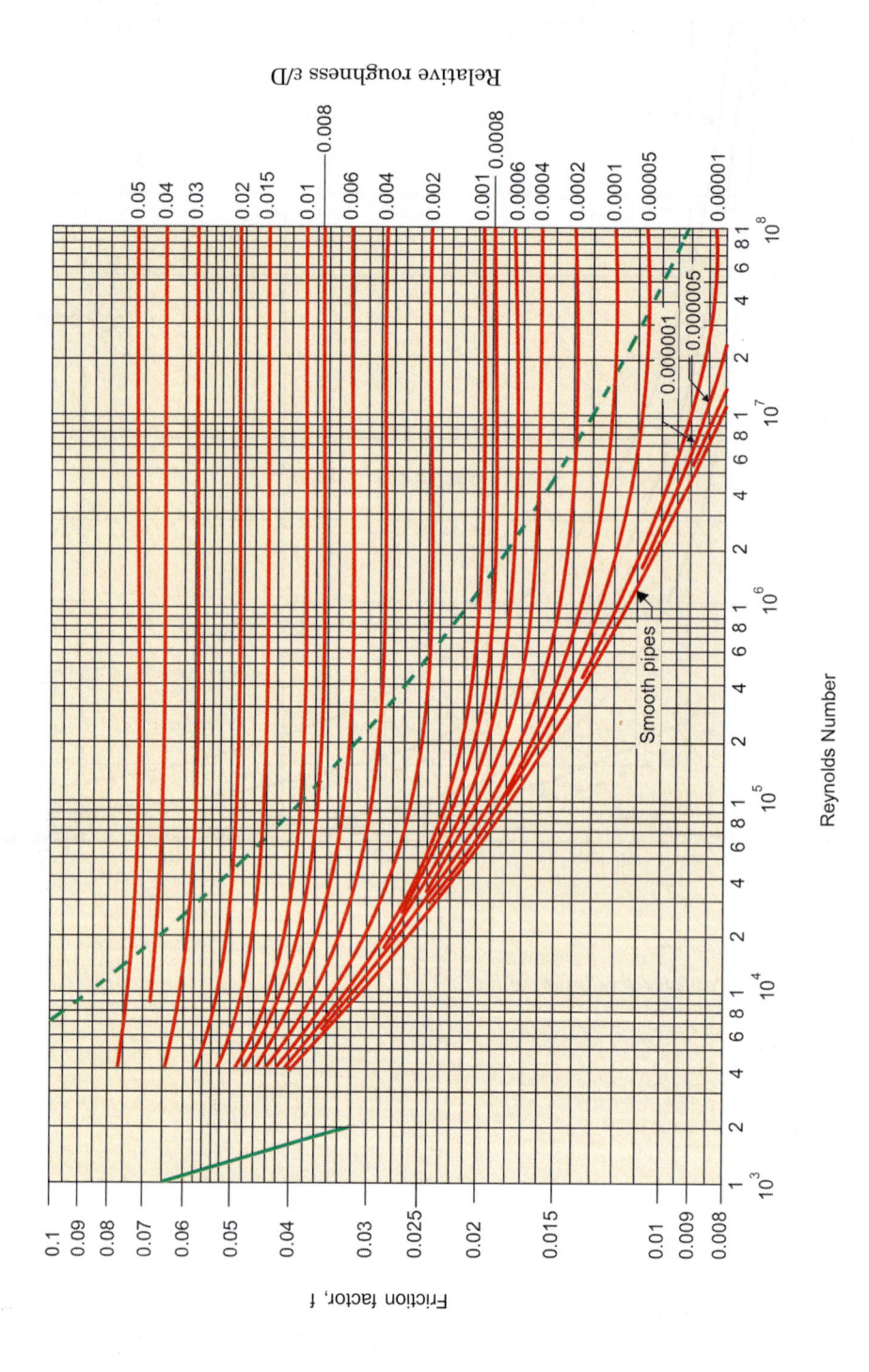

HEAT AND MASS TRANSFER DATA BOOK

PACKED BEDS

Author	Correlation	Notations
ERGUN	LIQUIDS: $$Re_p = \frac{DU_{bs}\rho}{\mu(1-\varepsilon)}$$	Re_p — Reynolds Number of packed bed
ERGUN	$$f_p = \frac{g\,DH_f\varepsilon^3}{LU_{bs}^2(1-\varepsilon)}$$ also	D — effective particle diameter $= \dfrac{6}{S_v}$
KOZENY-CARMAN	$f_p = \dfrac{150}{Re_p}$ for $Re_p < 1$	S_v — specific surface of a particle $= S_p/V_p$
BURKE-PLUMMER	$f_p = 1.75$ for $Re_p > 2500$	S_p — surface area of particle
ERGUN	$f_p = \dfrac{150}{Re_p} + 1.75$ for $1 < Re_p < 2500$ $S_{vm} = \Sigma x_i S_{vi}$ $D_m = \dfrac{6}{S_{vm}} = \dfrac{1}{\Sigma(x_i/D_i)}$	V_p — volume of a particle S_{vm} — mean specific surface U_{bs} — superficial velocity, based on the area of an equivalent empty container $= \varepsilon U_b$ U_b — average interstitial velocity of fluid ε — void fraction or porosity
BEEK	$$f_p = \frac{1-\varepsilon}{\varepsilon^3}\left(1.75 + 150\frac{1-\varepsilon}{Re_p}\right)$$	ρ — density of the fluid μ — absolute viscosity of the fluid g — gravitational constant H_f — head of fluid lost due to friction L — length of packed bed f_p — friction factor x_i — volume fraction of particles of same size D_i — effective particle diameter of size i D_m — mean effective diameter

(Contd.)

PACKED BEDS *(Contd.)*

Author	Correlation	Notations	
	GASES : Pressure drop, $\Delta P = \dfrac{2f_p L G^2}{D \rho_{av}}$ for $\dfrac{\Delta p}{p} < 0.1$	G ρ_{av} p	mass velocity of the gas per unit area, kg/m^2 density of gas at the arithmetic mean pressure pressure of gas at entry
ECKERT WHITAKER BEEK	**HEAT TRANSFER :** $Nu_p = \dfrac{hD_p}{k} = 0.8\, Re_p^{0.7}\, Pr^{0.333}$ $\dfrac{hD_p}{k} =$ $\dfrac{1-\varepsilon}{\varepsilon}[0.5\, Re_p^{0.5} + 0.2\, Re_p^{0.67}]\, Pr^{1/3}$ From wall to gas, cylinder like filling $\dfrac{hD_p}{k} = 2.58\, Re_p^{0.33}\, Pr^{0.33} +$ $\quad 0.094\, Re_p^{0.8}\, Pr^{0.4}$ For sphere like filling $\dfrac{hD_p}{k} = 0.203\, Re_p^{0.33}\, Pr^{0.33}$ $\quad\quad + 0.22\, Re_p^{0.8}\, Pr^{0.4}$ $40 < Re_p < 2000$	St h Nu_p D_p k j_d h_D Sc	Stanton number heat transfer coeff. Nusselt Number for packed bed effective particle diameter thermal conductivity of fluid COLBURN j factor $= \dfrac{h_D}{u_\infty} Sc^{0.667}$ convective mass transfer coefficient, m/s Schmidt number
UPADHYAY	$St = \dfrac{1}{\varepsilon}\, 1.075\, Re_p^{-0.826}$ $0.01 < Re_p < 10$ $St = \dfrac{1}{\varepsilon}\, 0.455\, Re_p^{-0.4}$ $10 < Re_p < 200$		
	MASS TRANSFER : $j_d = 1.82\, Re_p^{-0.51}$ for $Re_p < 350$ $\quad = 0.989\, Re_p^{-0.41}$ for $Re_p > 350$		

HEAT AND MASS TRANSFER DATA BOOK

FREE CONVECTION

Note: Properties are to be evaluated at $T_f = (T_w + T_\infty)/2$ unless otherwise mentioned.

Description	Correlation and Validity	Notations
1.1 Vertical Surface : Plate or cylinder with $D \geq \dfrac{35}{Gr_L^{0.25}}$ Laminar : $Gr \cdot Pr < 10^9$	$\delta_x = 3.93\, x\, Pr^{-0.5}\, (0.952 + Pr)^{0.25} \cdot Gr_x^{-0.25}$ $h_x = \dfrac{2k}{\delta_x}$; $Nu_x = 2\dfrac{x}{\delta_x}$ constant wall temperature $Nu_x = 0.508\, Pr^{0.5}\, (0.952 + Pr)^{-0.25} \cdot Gr_x^{0.25}$ for constant heat flux $Nu_x = 0.6\, (Gr_x\, Nu_x\, Pr)^{0.2}$ for $10^5 < Gr_x\, Nu_x < 10^{11}$ $Nu_x = 0.6\, (Gr^* Pr)^{0.2}$ $Nu_x = C(Gr_x)^{0.25}$ where C is a function of Prandtl number as tabulated below : Both constant wall temperature and constant heat flux | Pr	0.01	0.72	1.0	2.0	10	100	1000 | |----|----|----|----|----|----|----|----| | C | 0.057 | 0.357 | 0.401 | 0.506 | 0.827 | 1.549 | 2.804 |	T_f — film temperature T_w — wall temperature T_∞ — free stream fluid temp. δ_x — boundary layer thickness at x from leading edge $Gr_x = \dfrac{g \cdot \beta \cdot x^3\, \Delta T}{v^2}$ Nu_x — local Nusselt Number $\overline{Nu}$ — average Nusselt Number upto length L Nu_L — Nusselt Number at L from leading edge k — Thermal conductivity of fluid v — kinematic viscosity of fluid $Gr^* = \dfrac{g\beta q\, x^4}{kv^2}$ q — heat flux, W/m²
Average Nusselt Number upto L	$\overline{Nu} = \dfrac{4}{3} Nu_L$	

*For inclined plates replace Gr_x by $Gr_x \cos\theta$, where θ is angle with vertical.

Description	Correlation and Validity	Notations
1.2 For Low Values of $Gr\,Pr < 10^4$	$\overline{Nu} = 0.68 + \dfrac{0.67\,(GrPr)^{0.25}}{\left\{1 + \left[\dfrac{0.492}{Pr}\right]^{0.5625}\right\}^{0.444}}$ Constant wall temperature For constant heat flux, use 0.437 in place of 0.492	θ, angle with vertical
Higher values of $Gr\,Pr$ Constant heat flux/ constant wall temperature	$\overline{Nu} = 0.59\,(GrPr)^{0.25}$ $10^4 < GrPr < 10^9$	
All values of $Gr\,Pr$	$Nu = \left[0.825 + \dfrac{0.387\,(Gr\,Pr)^{0.167}}{\left\{1 + \left[\dfrac{0.492}{Pr}\right]^{0.5625}\right\}^{0.296}}\right]^2$ for Constant wall temperature For constant heat flux, use 0.437 in place of 0.492	
Inclined plates upto 60° with vertical	Multiply Gr by $\cos\theta$	
2.1 **TURBULENT** $GrPr > 10^9$ Both constant heat flux and constant wall temperature for constant heat flux	$Nu = 0.10\,(GrPr)^{0.333}$ (more popularly used) OR $Nu = 0.021\,(GrPr)^{0.4}$ $10^9 > GrPr < 10^{13}$ $Nu_x = 0.17\,(Gr_x Nu_x Pr)^{0.25}$ for $2\times 10^{13} < Gr_x Nu_x Pr < 10^{16}$ Also $Nu_x = 0.17\,(Gr^* Pr)^{0.25}$ $\overline{Nu} = \dfrac{5}{4} Nu_L$	$\overline{Nu}$ — Average Nu upto L $Gr^* = \dfrac{g\beta q x^4}{k v^2}$
2.1.1 **Horizontal plate:** upper surface heated or lower surface cooled constant wall temperature	Gr to be calculated with $L =$ Area/Perimeter $Nu = 0.54\,(GrPr)^{0.25}$; $2\times 10^4 < GrPr < 8\times 10^6$ $Nu = 0.15\,(GrPr)^{0.333}$; $8\times 10^6 < GrPr < 10^{11}$	q — heat flux, W/m^2

HEAT AND MASS TRANSFER DATA BOOK

Description	Correlation and Validity	Notations
2.1.2 Upper surface cooled or lower surface heated Constant wall temperature	$Nu = 0.27 (GrPr)^{0.25}$ $10^5 < GrPr < 10^{11}$	Nu = average Nusselt number
2.2.1 Horizontal Plate: Heated Surface up Constant heat flux	Properties except β to be evaluated at $T_e = T_w - 0.25 (T_w - T_\infty)$ $Nu = 0.13 (Gr_e Pr_e)^{0.333}$; $Gr_e Pr_e < 2 \times 10^8$ $Nu = 0.16 (Gr_e Pr_e)^{0.333}$; $2 \times 10^8 < Gr_e Pr_e < 10^{11}$	Gr_e = Grashof with properties evaluated at T_e $T_e = T_w - 0.25 (T_w - T_\infty)$ Pr_e = Prandtl at T_e β — evaluated at T_∞ T_w — Average wall temp.
2.2.2 Heated surface down constant heat flux	$Nu = 0.58 (Gr_e . Pr_e)^{0.2}$ $10^6 < Gr_e Pr_e < 10^{11}$	
2.3.1 Inclined Plates Heated surface facing down constant heat flux	$Nu = 0.56 (Gr_e Pr_e \cos \theta)^{0.25}$ $\theta < 88°$, $10^5 < Gr_e Pr_e \cos \theta < 10^{11}$ $Nu_x = 0.17 (Gr_x^* . Nu_x . Pr)^{0.25}$ $10^{10} < Gr_x^* Nu_x . Pr < 10^{15}$ for heated face facing down, multiply Gr^* by $\cos^2 \theta$, $Nu = 0.58 (Gr_e Pr_e)^{0.2}$ $88° < \theta < 90°$ $10^6 < GrPr < 10^{11}$	θ = angle with vertical $Gr^* = \dfrac{g\beta q x^4}{k\nu^2}$ q = heat flux, W/m^2
2.3.2 Heated Surface up Constant heat flux	$Nu = 0.14 [(Gr_e Pr_e)^{0.333} - (A Pr_e)^{0.333}]$ $+ 0.56 (Gr_e Pr_e \cos \theta)^{0.25}$ $10^5 < Gr_e Pr_e \cos \theta < 10^{11}$ A is found from the tabulation below: | θ | 15 | 30 | 60 | 75 | |---|---|---|---|---| | A | 5×10^9 | 2×10^9 | 10^8 | 10^6 |	$Z = 0.25 + 0.083 (\sin \theta)^{1.75}$
2.4 Inclined Cylinders constant heat flux	$Nu_L = [0.60 - 0.488 (\sin \theta)^{1.03}] (Gr_L Pr)^Z$ $Gr_L Pr < 2 \times 10^8$,	Gr_L—Based on length of cylinders

Description	Correlation and Validity	Notations
3.1 Horizontal cylinders (long cylinders)	$Nu = C\,(Gr_D Pr)^m$ (constant wall temp.) $\begin{array}{ccc} Gr_D Pr & C & m \\ 10^{-10} \text{ to } 10^{-2} & 0.675 & 0.058 \\ 10^{-2} \text{ to } 10^{2} & 1.02 & 0.148 \\ 10^{2} \text{ to } 10^{4} & 0.85 & 0.188 \\ 10^{4} \text{ to } 10^{7} & 0.48 & 0.25 \\ 10^{7} \text{ to } 10^{12} & 0.125 & 0.333 \end{array}$	
3.2 Horizontal Cylinder	$Nu_D = \left\{ 0.60 + 0.387 \left[\dfrac{Gr_D Pr}{\left\{1 + \left(\dfrac{0.559}{Pr}\right)^{0.5625}\right\}^{0.296}} \right]^{0.167} \right\}^2$ $10^{-5} < Gr_D Pr < 10^{12}$ $Nu_D = 0.36 + \dfrac{0.518\,(Gr_D Pr)^{0.25}}{\left[1 + \left(\dfrac{0.559}{Pr}\right)^{0.5625}\right]^{0.444}}$ $10^{-6} < Gr_D Pr < 10^{9}$	Gr_D — based on diameter
3.3 Liquid Metals	$Nu_D = 0.53\,[Gr_D (Pr)^2]^{0.25}$	
4.1 Spheres	$Nu = 2 + 0.43\,(Gr_D Pr)^{0.25}$ $1 < Gr_D Pr < 10^5$; $Pr \simeq 1$ $Nu = 2 + 0.50\,(Gr_D Pr)^{0.25}$ $3 \times 10^5 < Gr_D Pr < 8 \times 10^8$	α = Thermal diffusivity L = Refer figure
5.1 Rectangular Cavities (Enclosed space)	Surfaces of height H displaced by distance L and inclined at $\theta°$ to horizontal, surfaces at T_1 and T_2 $Q = hA\,[T_1 - T_2]$ $Nu_L = 0.069\,Ra_L^{0.333} \cdot Pr^{0.074}$ $3 \times 10^5 < Ra_L < 7 \times 10^9$, $\theta = 0°$	Properties at $T = (T_1 + T_2)/2$ $A = H \times$ length of cavity $Ra_L = \dfrac{g \cdot \beta\,(T_1 - T_2)\,L^3}{\alpha \nu}$ Q = heat transferred

HEAT AND MASS TRANSFER DATA BOOK

Description	Correlation and Validity	Notations
5.2 Medium aspect ratio (H/L)	$Nu_L = 0.22 \left(\dfrac{Pr}{0.2 + Pr} \cdot Ra_L \right)^{0.28} (H/L)^{-0.25}$ $Pr < 10,\ 2 < \dfrac{H}{L} < 10,\ Ra_L < 10^{10},\ \theta = 90°$	
5.3 Small aspect ratio (H/L)	$Nu_L = 0.18 \left(\dfrac{Pr}{0.2 + Pr} \cdot Ra_L \right)^{0.29}$ $1 < \dfrac{H}{L} < 2,\ 10^{-3} < Pr < 10$ $10^3 < [Ra_L \cdot Pr/(0.2 + Pr)]\ ;\ \theta = 90°$	
5.4 Large aspect ratio, lower Ra_L	$Nu_L = 0.42\, Ra_L^{0.25}\, Pr^{0.012} \left(\dfrac{H}{L} \right)^{-0.3}$ $10 < \dfrac{H}{L} < 40,$ $10^4 < Ra_L < 10^7$ $1 < Pr\quad \theta = 90°$	$Ra_L = \dfrac{g\beta (T_1 - T_2) L^3}{\alpha \nu}$ α = Thermal diffusivity
5.5 Large aspect ratio, higher Ra_L	$Nu_L = 0.046\, Ra_L^{0.333}$ $1 < \dfrac{H}{L} < 40\ ;\ 1 < Pr < 20,$ $10^6 < Ra_L < 10^9\quad \theta = 90°$	
5.6 Inclined cavities (below critical angle)	$Nu_L = 1 + 1.44 \left[1 - \dfrac{1708}{Ra_L \cdot \cos\theta} \right]^{@}$ $\times \left[1 - \dfrac{1708\, (\sin 1.8\theta)^{1.6}}{Ra_L \cdot \cos\theta} \right]$ $+ \left[\left(\dfrac{Ra_L \cos\theta}{5830} \right)^{0.333} - 1 \right]^{@}$ @ If the value of the expression is –ve set it to zero. $\dfrac{H}{L} \geq 12\quad 0° < \theta \leq \theta^*$ {| H/L	1	3	6	12	> 12 | | θ*	25	53	60	67	70 |}	θ^* = critical angle dependent on aspect ratio, refer table

H/L	1	3	6	12	> 12
θ*	25	53	60	67	70

Description	Correlation and Validity	Notations
5.7 Inclined cavities below critical angle θ^* small aspect ratio	$Nu_L = Nu_{L\theta=0} \left[\dfrac{Nu_{L\theta=90}}{Nu_{L\theta=0}} \right]^{\theta/\theta^*} (\sin \theta^*)^{\theta/4\theta^*}$ $\dfrac{H}{L} < 12 \;;\; 0 < \theta < \theta^*$	To calculate $Nu_{L\theta=0}$ and $Nu_{L\theta=90°}$ use equations in section 5.6.
5.8 Inclined cavities above critical angle	$Nu_L = Nu_{L(\theta=90)} (\sin \theta)^{0.25}$ $\theta^* < \theta < 90°$ $Nu_L = 1 + [Nu_{L(\theta=90)} - 1] \sin \theta$ $90° < \theta < 180°$	
6.1 Annular space between concentric cylinders	$q' = \dfrac{2\pi k_{eff}}{\ln\left(\dfrac{D_o}{D_i}\right)} [T_i - T_o]$ $\dfrac{k_{eff}}{k} = 0.386 \left(\dfrac{Pr}{0.861 + Pr} \right)^{0.25} (Ra_C)^{0.25}$ where $Ra_C = \dfrac{\left[\ln \dfrac{D_o}{D_i} \right]^4}{L^3 [D_i^{-0.6} + D_o^{-0.6}]^5} \cdot Ra_L$ $10^2 < Ra_C < 10^7$	k_{eff} = effective thermal conductivity, a stationary fluid should have to transfer the same amount of heat by conduction D_o = outside diameter D_i = inside diameter T_o = outside surface temperature T_i = inside surface temperature k = thermal conductivity of fluid L = $(D_o - D_i)/2$ Ra_L = Rayleigh number based on L
6.2 Space between concentric spheres	$q = k_{eff} \pi (D_i D_o / L) (T_i - T_o) = \dfrac{4\pi k_{eff} \cdot r_1 r_2 \, \Delta T}{r_2 - r_1}$ $\dfrac{k_{eff}}{k} = 0.74 \left(\dfrac{Pr}{0.861 + Pr} \right)^{0.25} (Ra_C)^{0.25}$ where $Ra_C = \left[\dfrac{L}{(D_o D_i)^4} \dfrac{Ra_L}{[D_i^{-1.4} + D_o^{-1.4}]^5} \right]^{0.25}$ $10^2 < Ra_C \le 10^4$	q' — heat transfer per unit length, W/m q — heat transfer between concentric spheres L — Gap length $= (r_2 - r_1)$

7.0 FREE CONVECTION: SIMPLIFIED EXPRESSIONS FOR CONVECTIVE HEAT TRANSFER COEFFICIENT FOR AIR AT ATMOSPHERIC PRESSURE

Geometry and Validity	h, W/m²K
7.1 LAMINAR ($10^4 < GrPr < 10^9$)	
Vertical plane or cylinder	$1.42 \left(\dfrac{\Delta T}{L}\right)^{0.25}$
Horizontal cylinder	$1.32 \left(\dfrac{\Delta T}{L}\right)^{0.25}$
Horizontal plate heated face up/cooled face down	$1.32 \left(\dfrac{\Delta T}{L}\right)^{0.25}$
(L = Area/Perimeter)	
Heated face down or cooled face up	$0.59 \left(\dfrac{\Delta T}{L}\right)^{0.25}$
7.2 TURBULENT	
$GrPr > 10^9$	
Vertical plane or cylinder	$1.31 (\Delta T)^{0.333}$
Horizontal cylinder	$1.24 (\Delta T)^{0.333}$
Horizontal plate:	
Heated face up or cooled face down	$1.52 (\Delta T)^{0.333}$
Heated face down or cooled face up.	$0.59 \left(\dfrac{\Delta T}{L}\right)^{0.25}$

For metric units divide the constant by 1.164

Note : For pressure other than atmospheric pressure multiply h by $\left(\dfrac{P}{\text{atm.pr}}\right)^{0.5}$ for laminar flow and by $\left(\dfrac{P}{\text{atm.pr}}\right)^{0.67}$ for turbulent flow. $\Delta T = T_w - T_\infty$

Geometry and Validity	Correlation	Notations
8.1 **COMBINED FREE AND FORCED CONVECTION**	$Gr/Re^2 \gg 1$ Free convection $Gr/Re^2 \approx 1$ Mixed convection $Gr/Re^2 \ll 1$ Forced convection	Gz = Graetz number $Gz = Re_D \cdot Pr \dfrac{D}{L}$
8.2 Mixed convection through tubes $(Gr/Re^2) \simeq 1$	**LAMINAR** $Nu = 1.75 \left[\dfrac{\mu}{\mu_w}\right]^{0.14} \cdot [Gz + 0.012(Gz \cdot Gr^{0.333})^{1.333}]^{0.333}$ **TURBULENT** $Nu = 4.69\, Re_D^{0.27}\, Pr^{0.21}\, Gr_D^{0.07}\, (D/L)^{0.36}$ $Re_D > 2000$ and $Ra_D \dfrac{D}{L} < 5000$ OR $Re_D > 800$ and $Ra_D \dfrac{D}{L} > 2 \times 10^4$	μ at bulk mean temp. μ_w at wall temp. $Re_w = \dfrac{\omega \pi D^2}{\nu}$ Peripheral speed Reynolds Number
9.1 Rotating Cylinder in a fluid	$Nu_D = 0.11\,[0.5\, Re_w^2 + Gr_D Pr]^{0.35}$ $Re_w > 8000$ in air	ω = rotational speed in rad/s
9.2 Rotating discs in a fluid	$Nu_D = \dfrac{\bar{h}D}{k} = 0.36 \left(\dfrac{\omega D^2}{\nu}\right)^{0.5}$ Rotational Reynolds No. $= \dfrac{\omega D^2}{\nu} < 5 \times 10^5$ $Nu_r = 0.0195\,(\omega R^2/\nu)^{0.8}$ $\dfrac{\omega D^2}{\nu} < 5 \times 10^5$ $Nu_r = \dfrac{\bar{h} r_o}{k} = 0.36 \left(\dfrac{\omega r_o^2}{\nu}\right)^{0.5} \left(\dfrac{r_c}{r_o}\right)^2$ $\qquad + 0.015 \left(\dfrac{\omega r_o^2}{\nu}\right)^{0.8} \left\{1 - \left(\dfrac{r_c}{r_o}\right)^{2.6}\right\}$	$D = 2r_o$ ν = kinematic viscosity of fluid Nu_r = local Nusselt Number based on radius R r_o = outside radius r_c = radius at which turbulence starts i.e. $\dfrac{\omega D_C^2}{\nu} = 5 \times 10^5$ $Re_D = \dfrac{\omega D^2}{\nu}$ $D_C = 2r_c$
9.3 Rotating Spheres	$Nu = 0.43\, Re_D^{0.5}\, Pr^{0.4}$ $Re_D < 5 \times 10^4$ $Nu_D = 0.066\, Re_D^{0.67}\, Pr^{0.4}$ $\qquad 5 \times 10^4 < Re_D < 7 \times 10^5$	

HEAT AND MASS TRANSFER DATA BOOK

BOILING

Description	Correlation and Validity	Notations
Nucleate Pool Boiling	$\dfrac{Q}{A} = \mu_l \cdot h_{fg} \left[g \dfrac{(\rho_l - \rho_v)}{\sigma} \right]^{0.5} \left[\dfrac{c_l \Delta T}{C_{sf} h_{fg} \text{Pr}^n} \right]^{3.0}$ Critical or maximum heat flux $\dfrac{Q}{A} = 0.18 h_{fg} \rho_v \left[\dfrac{\sigma \cdot g \cdot (\rho_l - \rho_v)}{\rho_v^2} \right]^{0.25}$ properties at $(T_w + T_{sat})/2$ $\dfrac{Q}{A} = 1.464 \times 10^{-9} \left[\dfrac{c_1 k_l^2 (\rho_l - \rho_v) \rho_l}{\mu_l^{0.5} \rho_v h_{fg}} \right]^{0.5} \left[\dfrac{h_{fg} \cdot \rho_v \Delta T}{\sigma T_l} \right]^{2.3}$ $\dfrac{Q}{A} = C \left[\dfrac{c_1 k_l^2 h_{fg}^3 (\rho_l - \rho_v) \rho_l^{0.5} \rho_v^3}{\mu_l^{0.5} \sigma^4 T_l^4} \right]^{0.5} (\Delta T)^{3.5}$ for C refer page 145	n = 1 for water and 1.7 for other fluids μ_l = dynamic viscosity of liquid μ_v = dynamic viscosity of vapour h_{fg} = enthalpy of evaporation, refer page 145, 148 g = acceleration due to gravity 9.81 m/s² ρ_l = density of liquid ρ_v = density of vapour σ = surface tension for liquid vapour interface. Refer p. 145, 147. c_l = specific heat of liquid ΔT = Excess temperature $T_w - T_{sat}$ T_w = surface temperature C_{sf} = constant, see tables on page 144 Pr_l = Prandtl Number for liquid Q/A = heat flux, W/m² T_l = liquid temp, K. k_l — Thermal conductivity of liquid k_v — Thermal conductivity of vapour c_{pv} — Specific heat of vapour at constant pressure σ_r — Stefan Boltzmann constant (see radiation) ε — Emissivity h — heat transfer coefficient in film boiling h_c — heat transfer coefficient due to convection h_r — heat transfer coefficient due to radiation T_w — wall temperature, K T_{sat} — saturation temperature of vapour, K D — diameter
Stable Film Boiling Horizontal tube Radiation prevails	$h_c = 0.62 \left[\dfrac{k_v^3 \rho_v (\rho_l - \rho_v) \cdot g \left(h_{fg} + 0.68 c_{pv} \Delta T \right)}{\mu_v D \Delta T} \right]^{0.25}$ $h_r = \sigma_r \varepsilon \left[\dfrac{T_w^4 - T_{sat}^4}{T_w - T_{sat}} \right]$ $h = h_c + 0.75 h_r$ Vapour properties at film temperature $\left(\dfrac{Q}{A}\right)_{\min} = 0.09 h_{fg} \cdot \rho_v \left[\dfrac{g \sigma (\rho_l - \rho_v)}{(\rho_l + \rho_v)^2} \right]^{0.25}$	

VALUES OF THE COEFFICIENT C_{sf} FOR VARIOUS LIQUID SURFACE COMBINATIONS

Liquid	Heating Surface	C_{sf}
Benzene	Chromium	0.010
Carbon Tetra Chloride	Copper	0.013
Carbon Tetra Chloride	Emery polished copper	0.007
Ethyl alcohol	Chromium	0.027
Isopropyl alcohol	Copper	0.00225
n – Butyl alcohol	Copper	0.00305
n – Pentane	Chromium	0.015
n – Pentane	Emery polished copper	0.0154
n – Pentane	Emery polished Nickel	0.0127
n – Pentane	Lapped copper	0.0049
n – Pentane	Emery rubbed copper	0.0074
Potassium carbonate K_2CO_3, 35%	Copper	0.0054
Potassium carbonate K_2CO_3, 50%	Copper	0.00275
Water	Copper	0.013
Water	Platinum	0.013
Water	Brass	0.0060
Water	Emery polished copper	0.0128
Water	Emery polished and paraffin treated copper	0.0147
Water	Scored copper	0.0068
Water	Teflon coated stainless Steel	0.0058
Water	Ground and polished stainless Steel	0.0080
Water	Chemically etched stainless Steel	0.0133
Water	Mechanically polished stainless Steel	0.0132

SIMPLIFIED EXPRESSIONS FOR BOILING HEAT TRANSFER COEFFICIENT FOR WATER AT ONE ATMOSPHERE*

Type of Surface	Range of Validity, W/m²	h, W/m²K	Notation
Horizontal	(Q/A) < 16000	$1042.6 \, (\Delta T)^{0.333}$	(Q/A) — Heat Flux, W/m²
	16000 < (Q/A) < 24000	$5.56 \, (\Delta T)^3$	h_p — Heat transfer coefficient at pressure, P
Vertical	(Q/A) < 3000	$537.3 \, (\Delta T)^{0.143}$	h — Heat transfer coefficient at atmospheric pressure, P_a
	3000 < (Q/A) < 63000	$7.96 \, (\Delta T)^3$	

*For other pressures, $h_p = h \times (P/P_a)^{0.4}$

C, CONSTANT FOR NUCLEATE BOILING, (SI Units)

Liquid	Surface	Apparatus		C
Water	Brass	Horizontal	Tube	7.6350×10^{-9}
Water	Brass Dirty	Horizontal	Plate	2.1400×10^{-9}
Water	Brass Clean	Horizontal	Plate	0.4890×10^{-9}
Water	Nickel	Horizontal	Wire	7.6350×10^{-9}
Water	Chromium	Horizontal	Tube	1.2200×10^{-9}
Water	Chromium	Horizontal	Plate	0.0460×10^{-9}
Acetone	Chromium	Horizontal	Plate	0.1070×10^{-9}
Ethanol	Copper	Horizontal	Plate	0.1830×10^{-9}
Ethanol	Chromium	Horizontal	Plate	0.0460×10^{-9}
n-Butanol	Brass	Horizontal	Tube	0.4890×10^{-9}
Propane	Copper	Horizontal	Tube	0.0915×10^{-9}
Propane	Chromium	Horizontal	Plate	0.0092×10^{-9}
n-Heptane	Chromium	Horizontal	Plate	0.0153×10^{-9}
CCl_4	Brass	Horizontal	Tube	0.6300×10^{-9}
CCl_4	Chromium	Vertical	Tube	0.0460×10^{-9}
Methyl Chloride	Copper	Horizontal	Tube	0.1373×10^{-9}

*LATENT HEAT OF VAPORISATION FOR WATER

Temp. °C	Latent Heat kJ/kg
0	2501.6
50	2382.9
100	2256.9
150	2113.2
200	1938.6
250	1714.6
300	1406.0
350	895.7
374.15	0.0

SURFACE TENSION, σ OF WATER AGAINST AIR

Temperature, °C	σ, N/m
0	0.0776
10	0.0742
20	0.0729
30	0.0711
40	0.0697
50	0.0679
60	0.0662
70	0.0644
80	0.0627
100	0.0588

*Also refer steam tables

RELATIVE MAGNITUDE OF HEAT TRANSFER COEFFICIENT IN NUCLEATE BOILING AT 1 atm RELATIVE TO VALUE FOR WATER

Fluid	h fluid/ h water
Water	1.0
Water 20% Sugar	0.87
10% Sodium sulphate, Na_2SO_4	0.94
26% Glycerin	0.83
55% Glycerin	0.75
24% Sodium Chloride, NaCl	0.61
Ispropanol	0.70
Methanol	0.53
Toluene	0.36
Carbon tetra Chloride	0.35
n-Butanol	0.32

APPROXIMATE BURNOUT HEAT VALUE AT 1 atm

Fluid-surface Combination	$(Q/A)max$ kW/m^2	ΔT critical
Water-copper	620–850	—
Water copper chrome plated	940–1260	23–28
Water steel	1290	30
Benzene-copper	130	—
Benzene-Aluminium	160	—
Propanol-Nickel plated copper	210–340	42–50
Butanol-Nickel plated copper	250–330	33–39
Ethanol-Aluminium	170	—
Ethanol-copper	250	—
Methanol-copper	390	—
Methanol-chrome plated copper	350	—
Methanol-steel	390	—
Liquid hydrogen	30	2
Liquid nitrogen	100	11
Liquid oxygen	150	11

SURFACE TENSION, σ OF WATER AGAINST ITS VAPOUR

Temperature, °C	σ, N/m
0	0.0755
20	0.0729
40	0.0695
60	0.0661
80	0.0627
100	0.0589
150	0.0487
160	0.0462
200	0.0378
250	0.0261
300	0.0143
350	0.0036
360	0.00156
374.1	0.0000

Approximately for water $\sigma = 0.1232 (1 - 0.001467\, T)$ N/m, T in K

SURFACE TENSION, σ OF LIQUIDS IN CONTACT WITH THEIR VAPOUR

Substance	Temperature, °C	σ, N/m
Acetic Acid	10	0.02890
Acetic Acid	20	0.02790
Acetic Acid	50	0.02480
Acetone	0	0.02630
Acetone	20	0.02370
Acetone	40	0.02120
Ammonia	11	0.02340
Ammonia	34	0.01820
Bromine	20	0.04150
Carbon Dioxide	20	0.00118
Carbon Dioxide	−10.55	0.00912
Carbon Tetrachloride	20	0.02700
Carbon Tetrachloride	100	0.01730
Carbon Tetrachloride	200	0.00660
Ethyl Alcohol	10	0.02360
Ethyl Alcohol	20	0.02270
Ethyl Alcohol	30	0.02190
Hydrazine	25	0.01710
Hydrogen Peroxide	18	0.01340
Methyl Alcohol	50	0.09180
Methyl Ether	−10	0.07640
Methyl Ether	−40	0.02010
Napthalene	127	0.02890
Tetrachlorethylene	20	0.03170
Toluene	10	0.02770
Toluene	20	0.02850
Toluene	30	0.02750

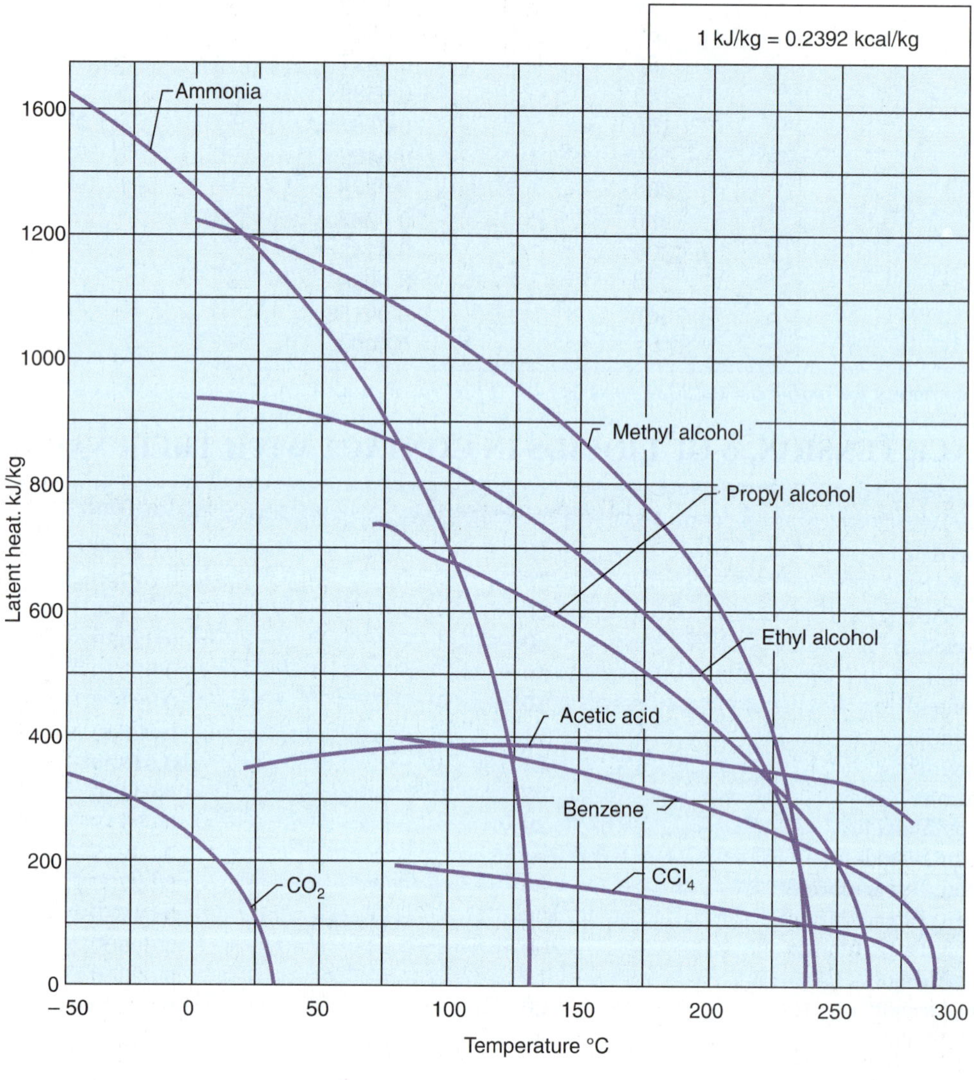
1 kJ/kg = 0.2392 kcal/kg
Ammonia
Methyl alcohol
Propyl alcohol
Ethyl alcohol
Acetic acid
Benzene
CO_2
CCl_4
Latent heat. kJ/kg
0
200
400
600
800
1000
1200
1600
– 50
0
50
100
150
200
250
300
Temperature °C

HEAT AND MASS TRANSFER DATA BOOK

CONDENSATION

$$Q = hA[T_v - T_s]$$

Note: Properties of Fluid are to be taken at $T_f = (T_v + T_s)/2$

Description	Correlation and Validity	Notations
FILM CONDENSATION : (Laminar) **(i) VERTICAL SURFACES**	$\delta_x = \left[\dfrac{4\mu_l\, kx\,(T_v - T_s)}{gh_{fg}\, \rho^2}\right]^{0.25}$ $\dot{m} = \dfrac{\rho_l(\rho_l - \rho_v)g\delta^3}{3\mu}$ $h_x = \dfrac{k}{\delta_x}$; $h = \dfrac{4}{3} h_L$ $h = 0.943 \left[\dfrac{k^3 \rho^2 g h_{fg}}{\mu_l L(T_v - T_s)}\right]^{0.25}$	δ_x = Boundary layer thickness at x, m μ_l = Dynamic viscosity of liquid k = Thermal conductivity of the liquid, W/mK x = Distance along the surface, m T_s = Surface temperature, T_v = Temperature of vapour g = Acceleration due to gravity, 9.81 m/s² h_{fg} = Latent heat of vaporisation, J/kg, refer page 148 ρ = Density of fluid, kg/m³ h_x = Local heat transfer coefficient h = Average heat transfer coefficient upto L, W/m²K $\dot{m}$ = Mass flow at any section L = Length of plate, m D = Diameter of plate, m N = Number of horizontal rows placed one above the other (irrespective of the number of tubes in a horizontal row) Re = Reynolds Number = $\dfrac{4\delta\rho u}{\mu}$ A = Flow area, m² p = shear perimeter, m u = average velocity of flow
(ii) HORIZONTAL TUBES	$h = 0.728 \left[\dfrac{k^3 \rho^2 g h_{fg}}{\mu_l D(T_v - T_s)}\right]^{0.25}$	
(iii) BANK OF N TUBES	$h = 0.728 \left[\dfrac{k^3 \rho^2 g h_{fg}}{\mu_l ND(T_v - T_s)}\right]^{0.25}$	
(iv) VERTICAL TUBES	$h = \dfrac{1.47}{\text{Re}^{0.33}} (k^3\rho^2 g/\mu^2)^{0.33}$ for Re < 1800 $h = 0.0077\, \text{Re}^{0.4} (k^3\rho^2 g/\mu^2)^{0.33}$ for Re > 1800	
(v) VERTICAL SHEET CONTAINING HORIZONTAL TUBES	$h = \dfrac{1.51}{\text{Re}^{0.33}}\left[\dfrac{k^3\rho^2 g}{\mu^2}\right]^{0.33}$ for Re < 1800	

CONDENSATION (Contd.)

Description	Correlation and Validity	Notations
(vi) INSIDE HORIZONTAL TUBES (Short tubes)	$h_D = 0.555 \left[\dfrac{g\rho_1(\rho_1 - \rho_v)k_1^3 h'_{fg}}{\mu_l(T_v - T_s)D} \right]^{0.25}$ $\dfrac{D \cdot G_v}{\mu_v} < 35{,}000$	G_v = mass flow rate of vapour per unit area G_l = mass flow rate of liquid per unit area c_l — Specific heat of liquid $h'_{fg} = h_{fg} + \dfrac{3}{8} c_l (T_{sat} - T_s)$ k_l, Pr_f at film temperature.
(vii) HIGHER VAPOUR VELOCITIES	$\dfrac{h_D D}{k_f} = 0.026\, Pr_f^{0.33}\, Re_m^{0.8}$ where $Re_m = \dfrac{D}{\mu_l}\left[G_l + G_v \left(\dfrac{\rho_l}{\rho_v}\right)^{0.5} \right]$ for $5000 < \dfrac{D \cdot G_v}{\mu_v} < 20{,}000$	

For inclined surfaces θ° with horizontal replace g by g sin θ

SIMPLIFIED EXPRESSIONS

Description	Correlation	Notations
(i) SINGLE VERTICAL TUBE	$h = 1.26\, Z \left[\dfrac{\pi d}{W}\right]^{-0.33}$	Z = condensation coefficient (Refer table on page 151) W = Weight of condensate in Newton/hr N_v = Number of vertical columns L = length of tube, **cm** N = total number of tubes d = diameter (outer) of tubes, **cm**
(ii) A BANK OF HORIZONTAL TUBES	$h = Z \left[\dfrac{N_v L}{W}\right]^{-0.33}$	
(iii) VERTICAL CONDENSERS	$h = 1.26\, Z \left[\dfrac{N\pi d}{W}\right]^{-0.33}$	

For inclined surfaces in place of g use g sin θ, angle θ with horizontal
or
g cos θ, angle θ with vertical.

CONDENSATION COEFFICIENT, Z, FOR FLUIDS. (SI Units)

Fluid	\multicolumn{6}{c}{Condensation Coefficient, z}
	0°C	50°C	100°C	150°C	200°C	300°C
Acetone	3125	3215	3182			
Ammonia	9860					
Benzene	2410	2801	2928	3187		
Deccine		1922	2042	2062	1922	
Dodeccine			1967	2019	2019	1785
Dowtherm–A					2398	2490
Ethyl Alcohol	1922	2398	2809			
Ethyl Glycol		2121	2639	2864		
Gasoline		2042	2104	2104		
Hexane	2156	2241	2258	2179		
Kerosene			1985	2042	1997	
Methyl Alcohol	2864	3389	3904			
Natural Gasoline	2121	2316	2428			
Octane	1985	2104	2196	2139		
Pentane	2333	2333	2258	2216		
i–Propyl alcohol		1780	2443			
Steam		11354	15413	19298		
Tetradeccine			1848	1942	1967	1820
Toluene		2667	2861			
Xylene		2876	2761			

For use with MKS units divide the above values of Z by 2.49

APPROXIMATE VALUES OF CONDENSATION HEAT TRANSFER COEFFICIENTS FOR VAPOURS AT 1 atm

Fluid	Geometry	W/m².K	$T_v - T_w$, °C
Steam	Vertical surface	4000–11,300	22–3
	Horizontal tubes, 15 to 76 mm diameter	9600–24,400	20–2
Diphenyl	Vertical surface, turbulent, 3.66 m	680–2400	72–13
	Horizontal tube, 43 mm diameter	1280–2270	15–5
Dowtherm A	Vertical surface 3.66 m turbulent	680–3060	40–20
Ethanol	Vertical surface, 152 mm	1130–2000	55–11
	Horizontal tube, 51 mm diameter	1800–2550	22–6
Propanol	Horizontal tube, 51 mm diameter	1400–1700	26–13
Butanol	Horizontal tube, 51 mm diameter	1400–1700	26–13
Benzene	Horizontal tubes, 15 to 33 mm diameter	1300–2150	45–13

T_v–Vapour temperature, T_w–Surface temperature

HEAT EXCHANGERS, $Q = UA \, (\Delta T)_{lm}$

Description	Correlation	Notations
SINGLE PASS (i) Parallel flow	$(\Delta T)_{lm} = \dfrac{(T_1 - t_1) - (T_2 - t_2)}{\ln\left[\dfrac{T_1 - t_1}{T_2 - t_2}\right]}$	Q — Heat exchanged, W U — Overall heat transfer coefficient W/m²K (Refer pages 156 and 157) $(\Delta T)_{lm}$ — Logarithmic mean temperature difference "LMTD" A — area, m² T_1 — Entry temperature of hot fluid T_2 — Exit temperature of hot fluid t_1 — Entry temperature of cold fluid t_2 — Exit temperature of cold fluid F — CORRECTION FACTOR depending on R, P and type of exchanger For F, R and P definition refer chart on pages 159 to 162 NTU — Number of Transfer Units = AU/C_{min} C_{min} — Smaller value of $m_h c_h$ and $m_c c_c$; W/K m_c — Mass flow rate of cold fluid, kg/s m_h — Mass flow rate of hot fluid c_h — Specific heat of cold fluid, J/kg K c_c — Specific heat of cold fluid, J/kg K ε — Effectiveness (depends on C_{min}/C_{max}) charts on pages, 163 to 168
(ii) Counter flow	$(\Delta T)^*_{lm} = \dfrac{(T_1 - t_2) - (T_2 - t_1)}{\ln\left[\dfrac{T_1 - t_2}{T_2 - t_1}\right]}$	
MULTIPASS and CROSS FLOW	$Q = FUA \, (\Delta T)^*_{lm}$	
NTU METHOD	$Q = \varepsilon C_{min}(T_1 - t_1)$ [Note. $\dfrac{C_{min}}{C_{max}}$ is zero when one of the fluids is condensing or evaporating and equal to 1 when $m_h c_h = m_c c_c$] $\varepsilon = \dfrac{m_h c_h}{C_{min}}\left[\dfrac{T_1 - T_2}{T_1 - t_1}\right]$ for $m_h C_h = C_{min}$ $= \dfrac{m_c c_c}{C_{min}}\left[\dfrac{t_2 - t_1}{T_1 - t_1}\right]$ for $m_c C_c = C_{min}$	

HEAT EXCHANGERS: EFFECTIVENESS RELATIONS

Description	Correlation	Notations
Double pipe Parallel flow	$\varepsilon = \dfrac{1 - \exp[-N(1+C)]}{1+C}$	$N = NTU = UA/C_{min}$
Counter flow	$\varepsilon = \dfrac{1 - \exp[-N(1-C)]}{1 - C\exp[-N(1-C)]}$	$C = \dfrac{C_{min}}{C_{max}}$
Counter flow $C = 1$	$\varepsilon = \dfrac{N}{N+1}$	$C_{min} = (mc)_{min}$
Cross flow Both fluids unmixed	$\varepsilon = 1 - \exp\left[\dfrac{\exp(-NC \cdot n) - 1}{Cn}\right]$	$n = N^{-0.22}$
Both fluids mixed	$\varepsilon = \left[\dfrac{1}{1-\exp(-N)} + \dfrac{C}{1-\exp(-NC)} - \dfrac{1}{N}\right]^{-1}$	ε, Effectiveness
C_{max} mixed and C_{min} unmixed	$\varepsilon = (1/C)\{1 - \exp[-C(1 - e^{-N})]\}$	m — flow rate kg/s
C_{max} unmixed and C_{min} mixed	$\varepsilon = 1 - \exp[-(1/C)(1 - \exp(-NC)]$	
Shell and tube : one shell pass and 2, 4, 6 tube passes	$\varepsilon = 2\left[1 + C + (1+C^2)^{0.5} \dfrac{1 + \exp[-N(1+C^2)^{0.5}]}{1 - \exp[-N(1+C^2)^{0.5}]}\right]^{-1}$	
All exchangers with $C = 0$	$\varepsilon = 1 - e^{-N}$	

HEAT EXCHANGERS: NTU RELATIONS

Description	Correlation	Notations
Double pipe parallel flow	$N = \dfrac{-\ln[1-(1+C)\varepsilon]}{1+C}$	N NTU
Counter flow	$N = \dfrac{1}{C-1} \ln\left(\dfrac{\varepsilon-1}{C\varepsilon-1}\right)$	$C = \dfrac{C_{min}}{C_{max}}$
Counter flow $C = 1$	$N = \dfrac{\varepsilon}{1-\varepsilon}$	ε effectiveness
Cross flow: C_{max} mixed and C_{min} unmixed	$N = -\ln\left[1 + \dfrac{1}{C} \cdot \ln(1-C\varepsilon)\right]$	expression may become indeterminate if $\Sigma(1+C) = 1$
C_{max} unmixed C_{min} mixed	$N = \dfrac{-1}{C}[1 + C \cdot \ln(1-\varepsilon)]$	
Shell and Tube One shell pass and 2, 4, 6 tube passes	$N = -(1+C^2)^{-0.5} \cdot \ln \dfrac{2/\varepsilon - 1 - C - (1+C^2)^{0.5}}{2/\varepsilon - 1 - C + (1+C^2)^{0.5}}$	$h = $ convective heat transfer coefficient A_L — heat transfer surface area for unit length of matrix
All exchangers $C = 0$	$N = -\ln(1-\varepsilon)$	c — specific heat of fluid
Storage type heat exchanger filled with matrix of solid material heated and cooled alternately.	Definitions: Location factor: at distance x $E = \dfrac{hA_L}{cm} x$ Time factor at time τ, $\eta = \dfrac{hA_L}{c_s M_L} \tau$	$\dot{m}$ — mass flow rate of fluid τ — time c_s — specific heat of matrix material M_L — mass of solid per unit length of matrix

(Contd.)

HEAT AND MASS TRANSFER DATA BOOK

Description	Correlation	Notations
Solid Temperature	$\dfrac{T - T_{go}}{T_o - T_{go}} = f_1 [E, \eta]$, refer chart on page 169	T_g — fluid temperature at location x, time τ
Gas Temperature	$\dfrac{T_g - T_{go}}{T_o - T_{go}} = f_2 [E, \eta]$, refer chart on page 169	T — matrix temperature at location x, time τ T_o — Initial temperature of matrix T_{go} — Entry temperature of fluid
Heat Transfer rate (Heating Period = cooling period)	$Q = \dfrac{1}{\dfrac{1}{h_H} + \dfrac{1}{h_C} + \dfrac{b}{3k}} \dfrac{A}{2} [T_{gH} - T_{gC}]$	Q — heat transfer per unit time h_H — convective heat transfer coefficient during heating h_C — Convective heat transfer coefficient during cooling k — thermal conductivity of matrix material
Effectiveness: (as per other heat exchangers)	Refer charts on page 170 Definitions for use of charts : $NTU_o = \dfrac{1}{C_{min}} \left[\dfrac{1}{\dfrac{1}{(hA)_H} + \dfrac{1}{(hA)_C}} \right]$ C_r = Matrix capacity rate = $c_s M_L N$ $(hA)^* = \dfrac{(hA)_C}{(hA)_H}$	A — Total material surface area T_{gh} — Average fluid temperature during heating period T_{gC} — Average fluid temperature during cooling period N — revolutions per hour.

APPROXIMATE OVERALL HEAT TRANSFER COEFFICIENTS FOR PRELIMINARY ESTIMATES

Duty	Overall Heat Transfer Coefficient, U W/m²K
Steam to water-instantaneous heater	2300–3500
Steam to water-storage-tank heater	1000–1800
Steam to heavy fuel oil	60–175
Steam to light fuel oil	175–350
Steam to light petroleum distillate	300–1200
Steam to aqueous solutions	600–3500
Steam to gases	30–300
Gas to gas	10–30
Water to compressed air	60–175
Water to water, jacket water coolers	900–1600
Water to lubricating oil	100–350
Water to condensing oil vapours	230–600
Water to condensing alcohol	250–700
Water to condensing Refrigerant-12	450–900
Water to condensing ammonia	900–1500
Water to organic solvents, alcohol	390–900
Water to boiling Refrigerant-12	300–900
Water to gasoline	350–550
Water to gas oil	200–350
Water to brine	600–1200
Light organics to light organics	250–450
Medium organics to medium organics	100–350
Heavy organics to heavy organics	60–250
Heavy organics to light organics	60–350
Crude oil to gas oil	175–350

FOULING RESISTANCE, R_f

Fluid	Fouling Resistance, R_f, m^2K/W
Sea Water below 52°C	0.0000827
Sea Water above 52°C	0.0017540
Treated Boiler Feed Water above 52°C	0.0001754
Fuel Oil	0.0008770
Quenching Oil	0.0007051
Alcohol vapours	0.0000877
Steam, not mixed with Oil	0.0000877
Industrial Air	0.0003525
Refrigerants	0.0001754

Dirt layers that gradually build up on heat transfer surfaces increase thermal resistance. This is accounted for by the use of Fouling Factors R_{fo} and R_{fi}.

EXPRESSIONS FOR OVERALL HEAT TRANSFER COEFFICIENTS, U_o AND U_i FOR TUBULAR SECTIONS

$$\frac{1}{U_o} = \frac{1}{h_o} + R_{fo} + \frac{r_o}{k}\ln\frac{r_o}{r_i} + \frac{r_o}{r_i}R_{fi} + \frac{r_o}{r_i}\frac{1}{h_i}$$

$$\frac{1}{U_i} = \frac{1}{h_i} + R_{fi} + \frac{r_i}{k}\ln\frac{r_o}{r_i} + \frac{r_i}{r_o}R_{fo} + \frac{r_i}{r_o}\frac{1}{h_o}$$

TEMA NOTATIONS FOR HEAT EXCHANGERS

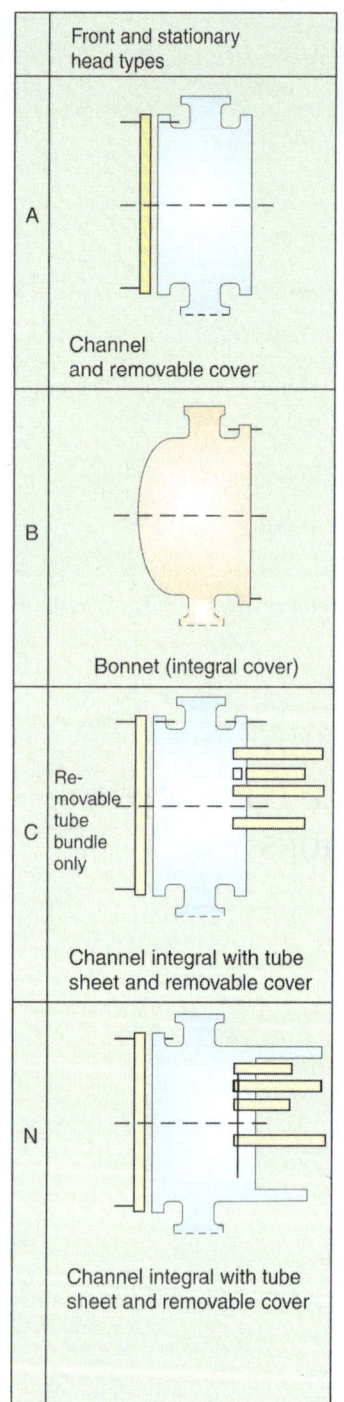

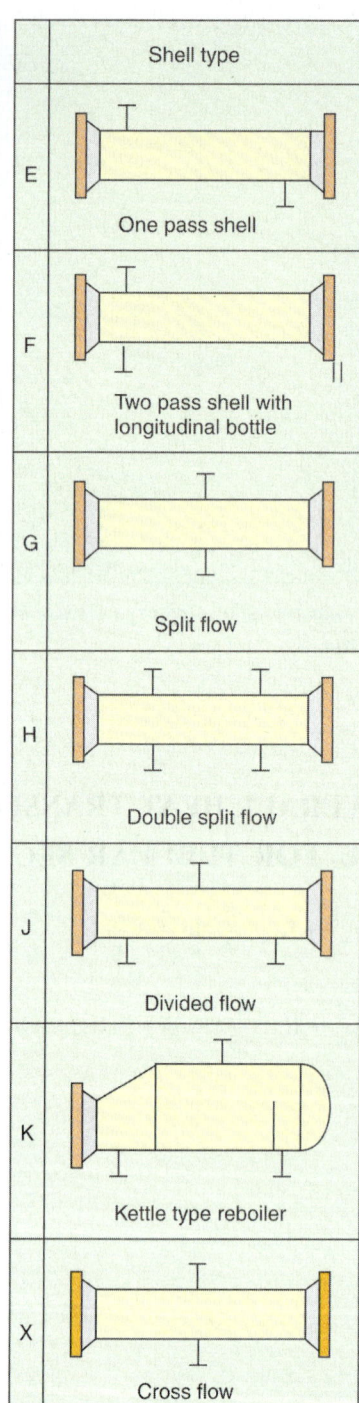

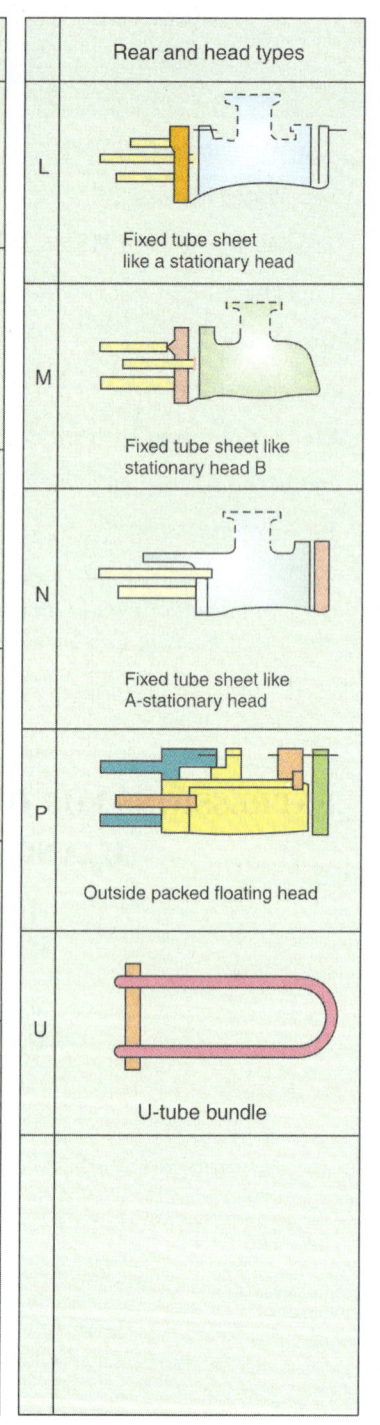

	Front and stationary head types		Shell type		Rear and head types
A	Channel and removable cover	E	One pass shell	L	Fixed tube sheet like a stationary head
B	Bonnet (integral cover)	F	Two pass shell with longitudinal bottle	M	Fixed tube sheet like stationary head B
C	Removable tube bundle only / Channel integral with tube sheet and removable cover	G	Split flow	N	Fixed tube sheet like A-stationary head
		H	Double split flow	P	Outside packed floating head
		J	Divided flow	U	U-tube bundle
N	Channel integral with tube sheet and removable cover	K	Kettle type reboiler		
		X	Cross flow		

CORRECTION FACTOR PLOT FOR EXCHANGER WITH ONE SHELL PASS AND TWO, FOUR OR MULTIPLE TUBE PASSES

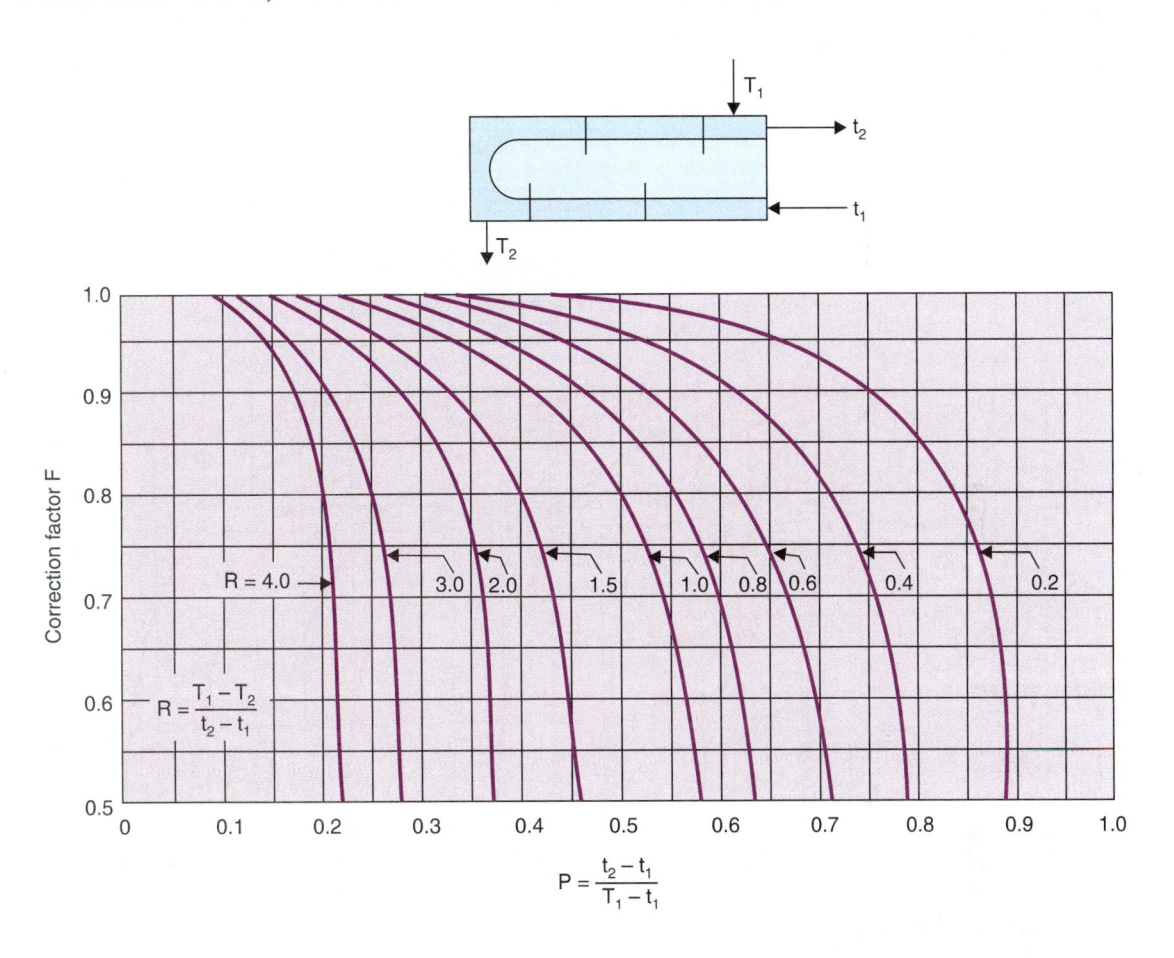

$$P = \frac{t_2 - t_1}{T_1 - t_1}$$

$$R = \frac{T_1 - T_2}{t_2 - t_1}$$

Correction factor F vs P for R = 4.0, 3.0, 2.0, 1.5, 1.0, 0.8, 0.6, 0.4, 0.2

CORRECTION FACTOR PLOT FOR EXCHANGER WITH TWO SHELL PASSES AND FOUR, EIGHT, OR ANY MULTIPLE NUMBER OF TUBE PASSES

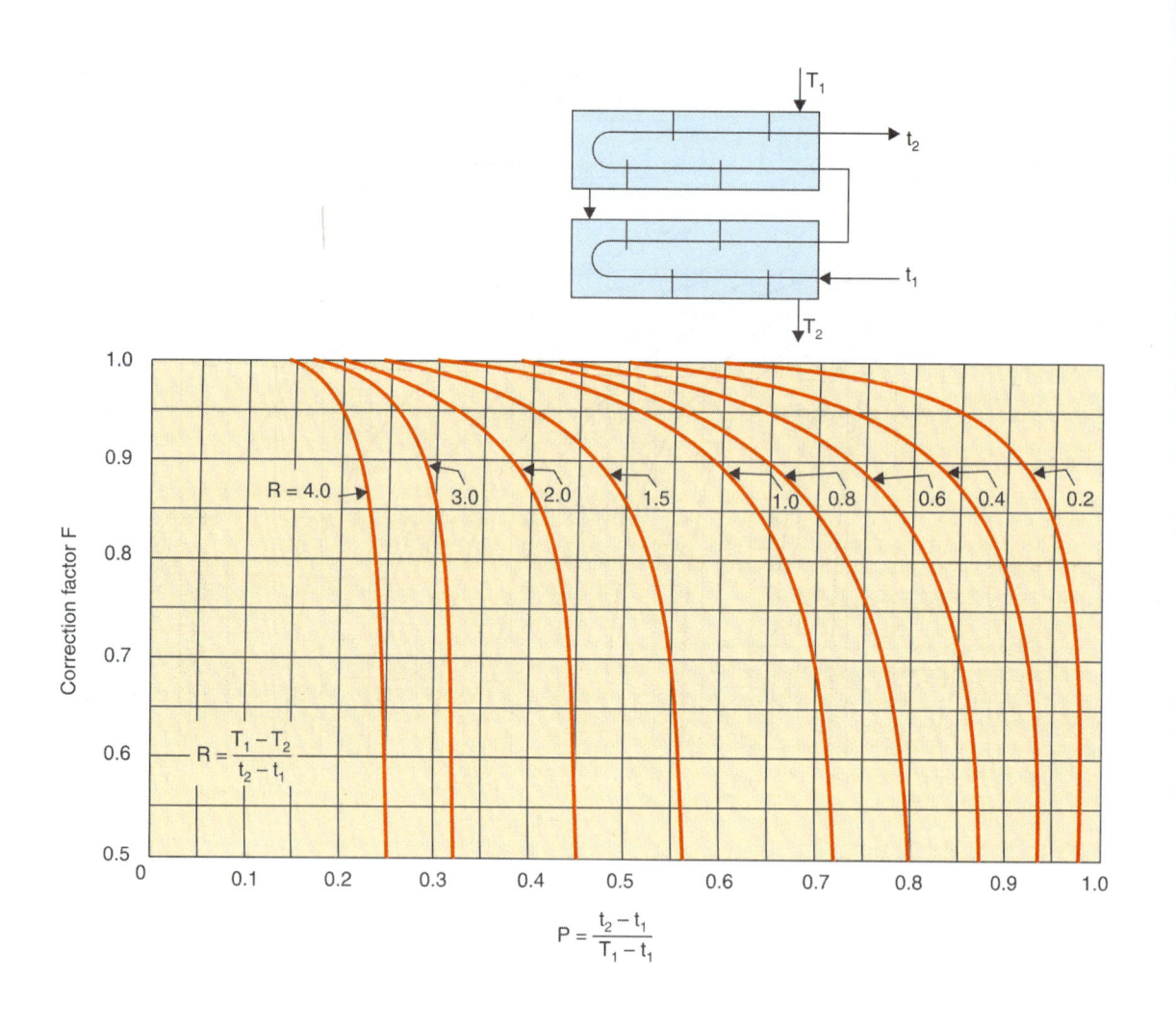

$$R = \frac{T_1 - T_2}{t_2 - t_1}$$

$$P = \frac{t_2 - t_1}{T_1 - t_1}$$

CORRECTION FACTOR PLOT FOR SINGLE PASS CROSS-FLOW EXCHANGER, ONE FLUID MIXED; OTHER UNMIXED

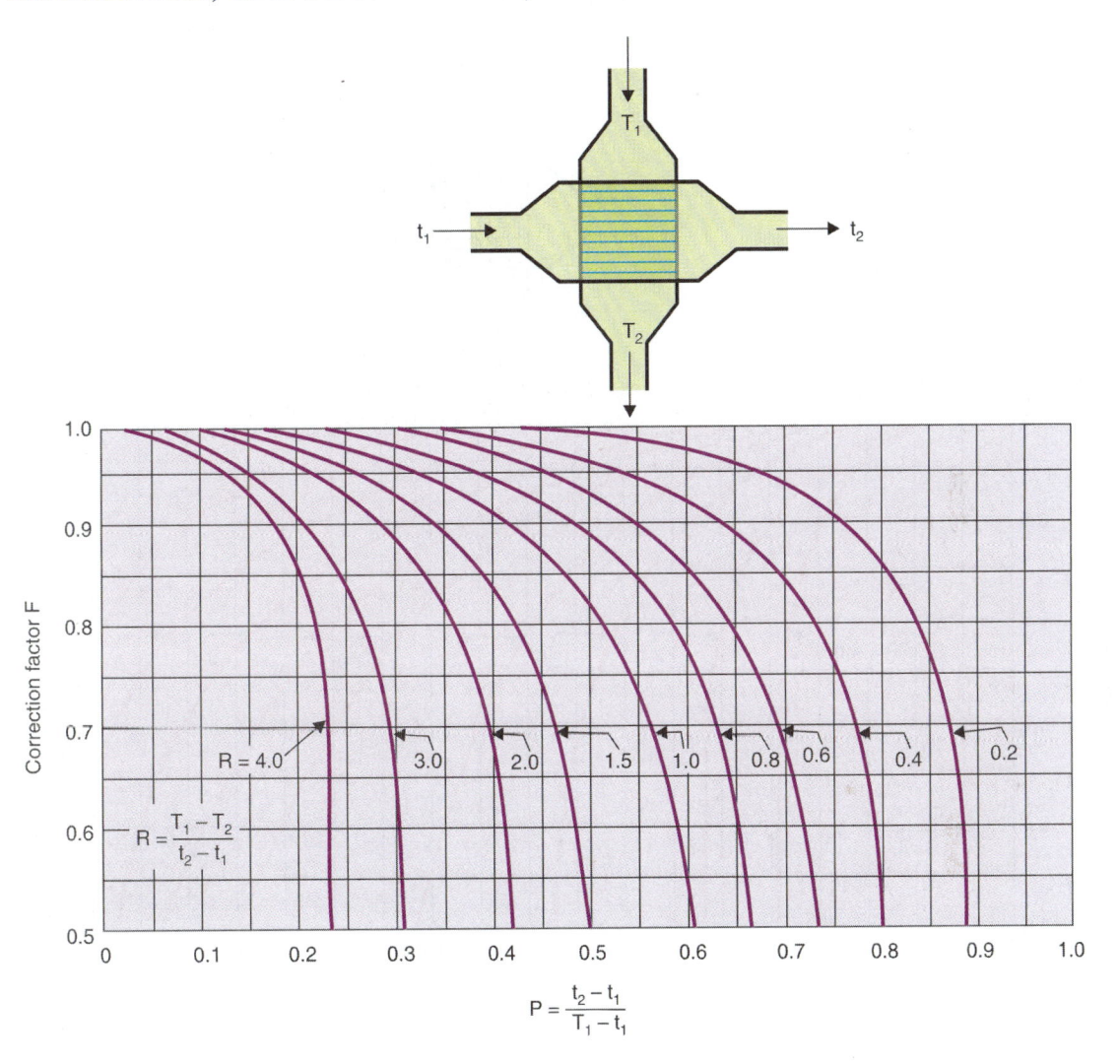

$$R = \frac{T_1 - T_2}{t_2 - t_1}$$

$$P = \frac{t_2 - t_1}{T_1 - t_1}$$

CORRECTION FACTOR PLOT FOR SINGLE PASS CROSS-FLOW EXCHANGER, BOTH FLUIDS UNMIXED

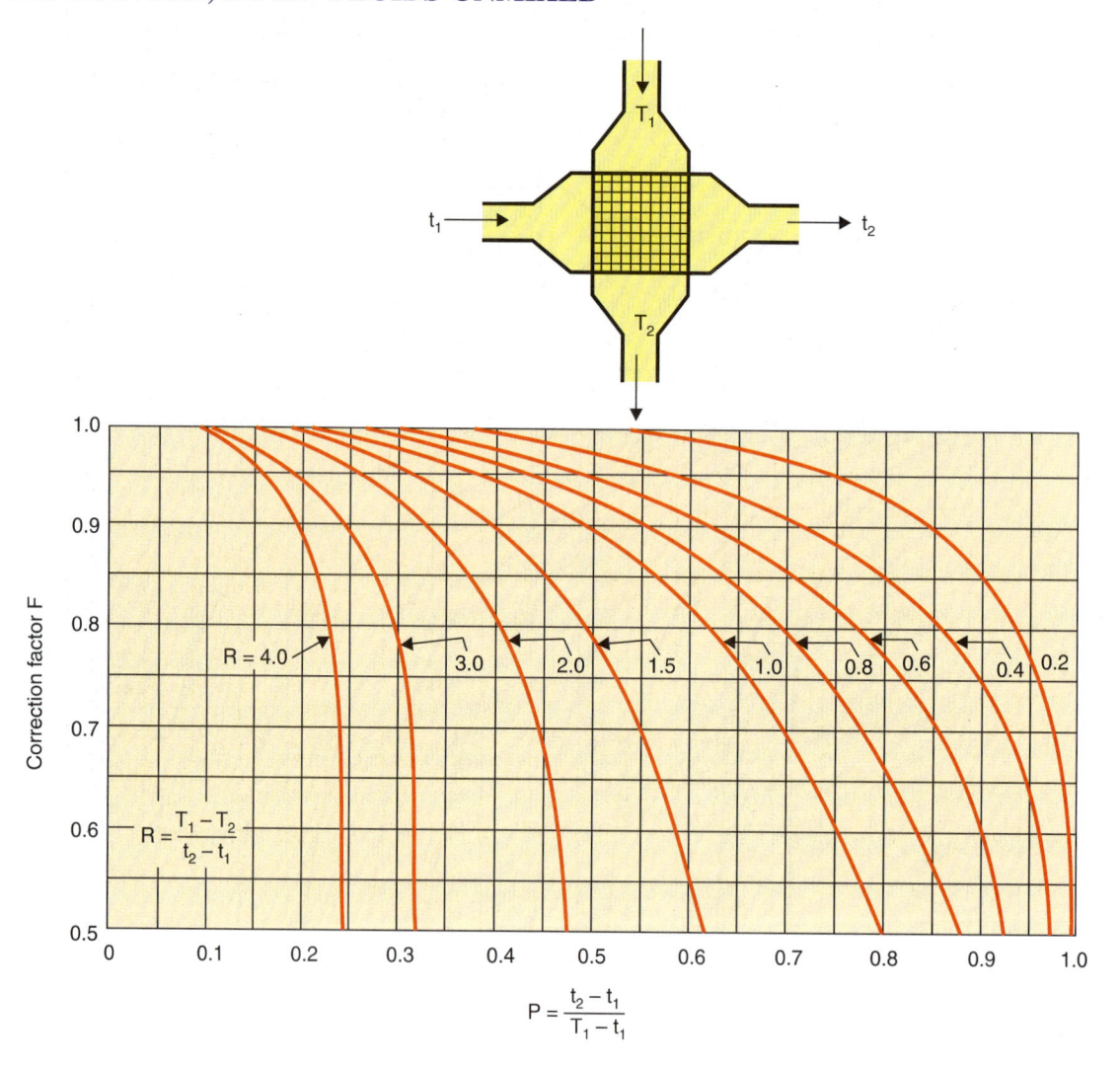

$$R = \frac{T_1 - T_2}{t_2 - t_1}$$

$$P = \frac{t_2 - t_1}{T_1 - t_1}$$

HEAT AND MASS TRANSFER DATA BOOK

EFFECTIVENESS—PARALLEL FLOW

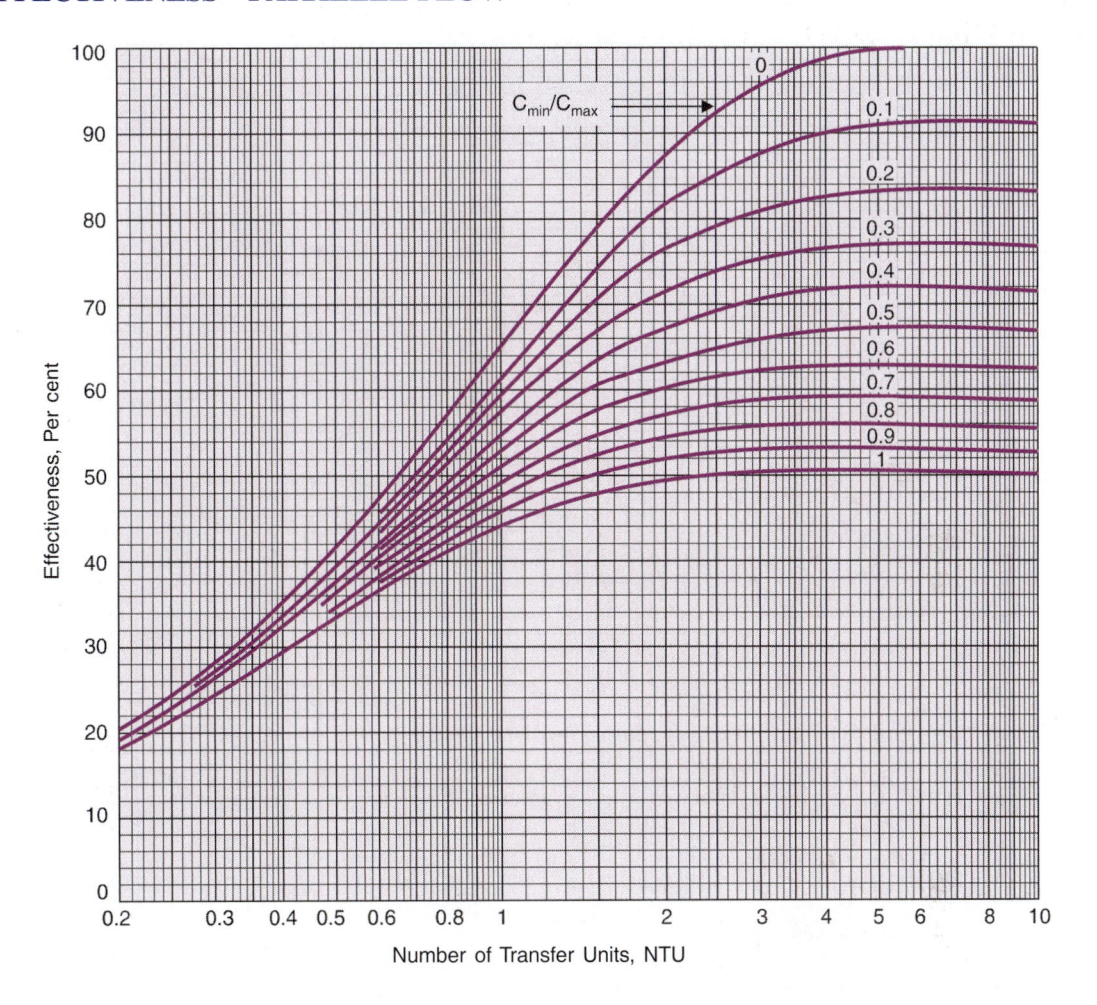

EFFECTIVENESS—COUNTER FLOW

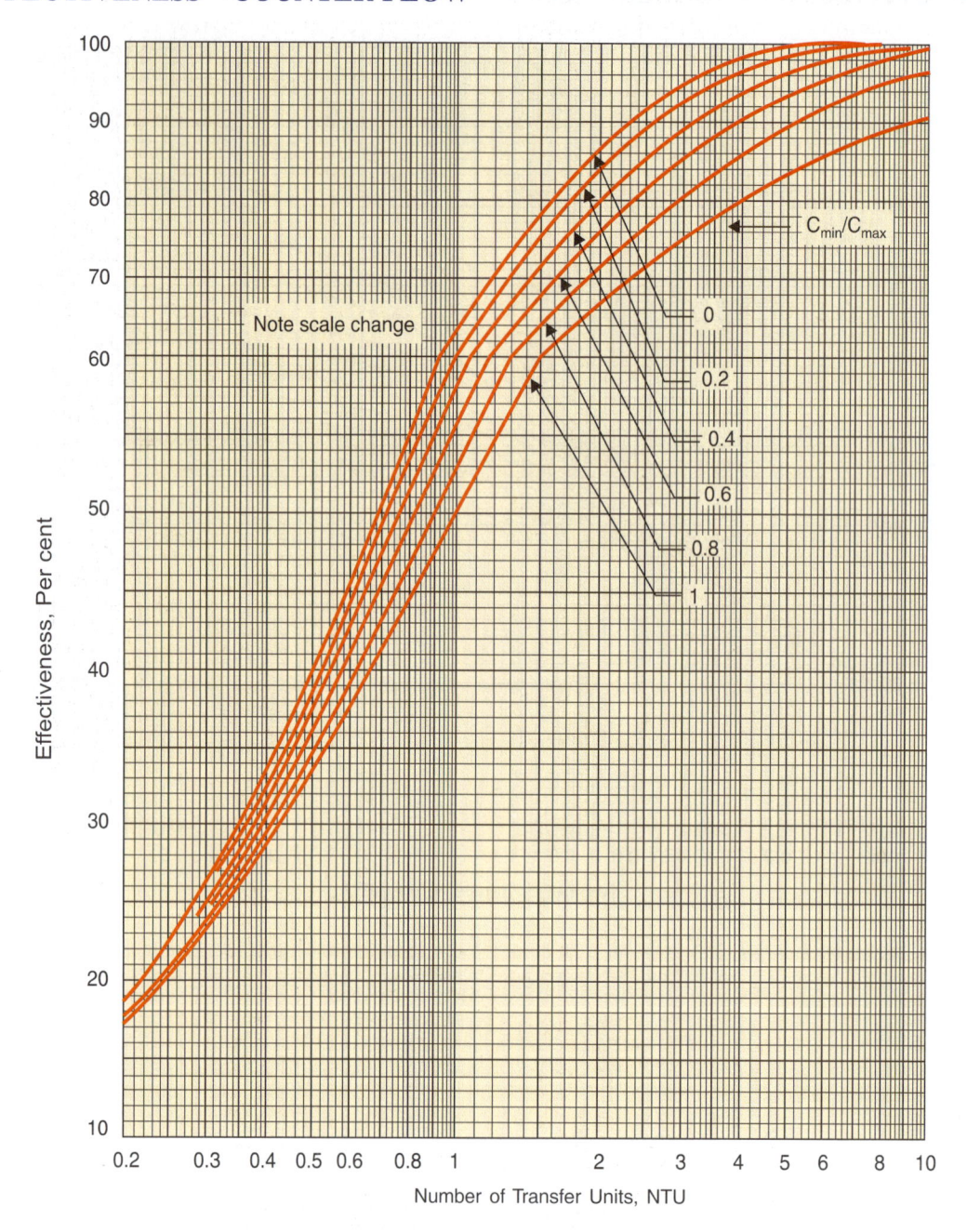

HEAT AND MASS TRANSFER DATA BOOK

EFFECTIVENESS—1 SHELL PASS, 2, 4, 6 TUBE PASSES

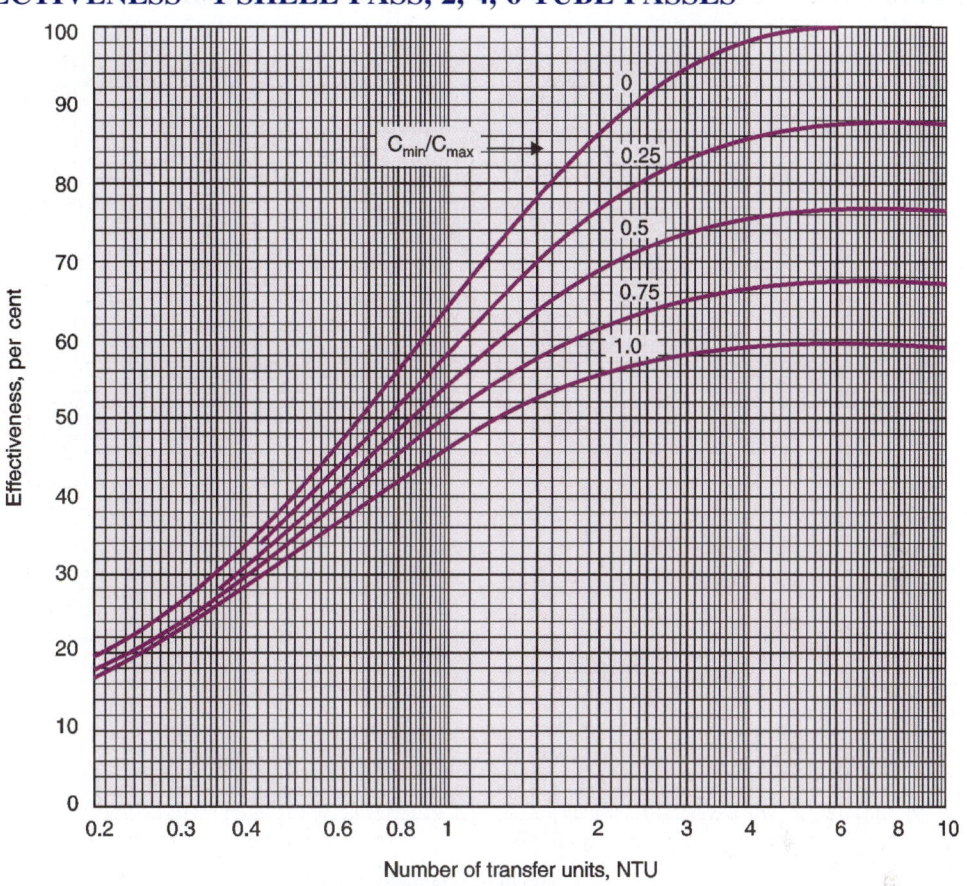

EFFECTIVENESS—2 SHELL, 4-8-12-TUBE PASSES

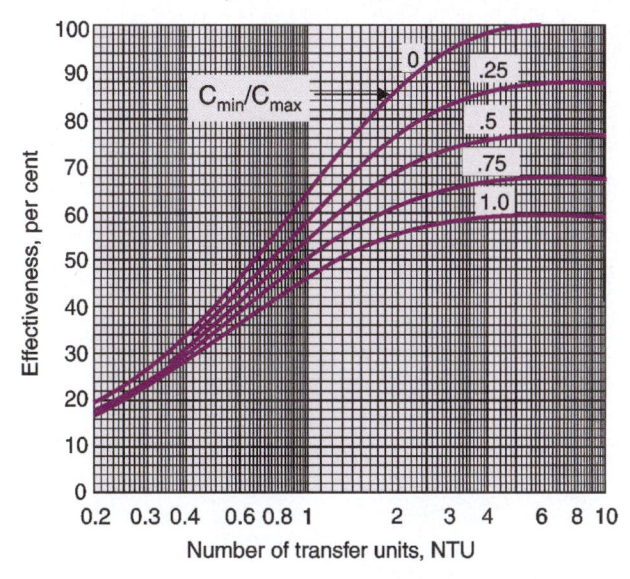

EFFECTIVENESS—CROSS FLOW, BOTH FLUIDS UNMIXED

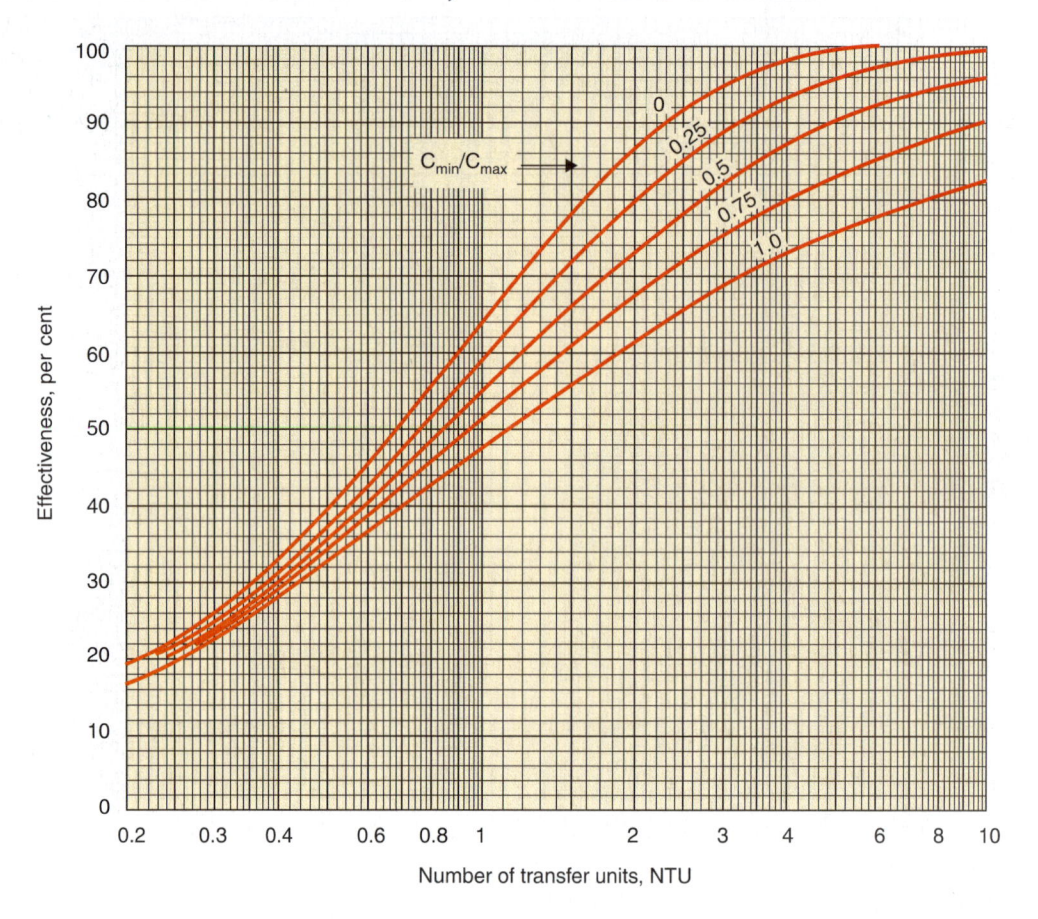

EFFECTIVENESS—CROSS FLOW, ONE FLUID MIXED

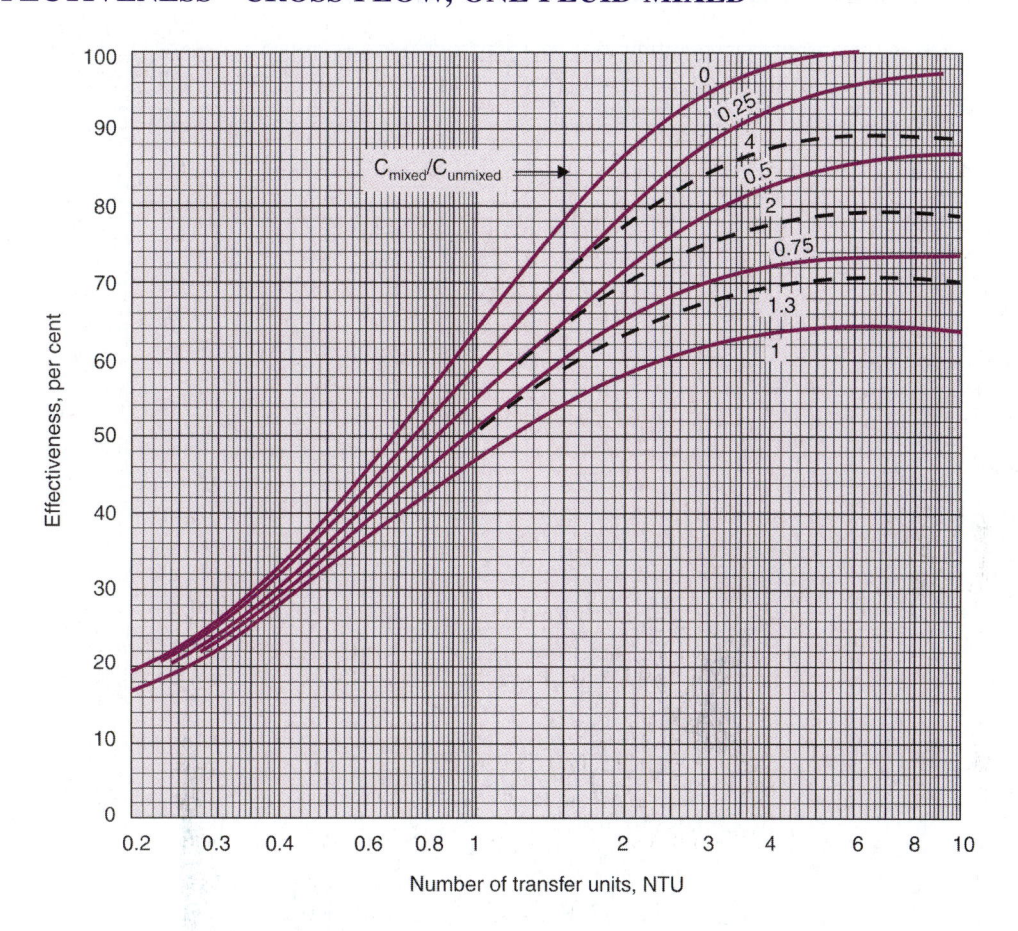

EFFECTIVENESS—CROSS FLOW—BOTH FLUIDS MIXED

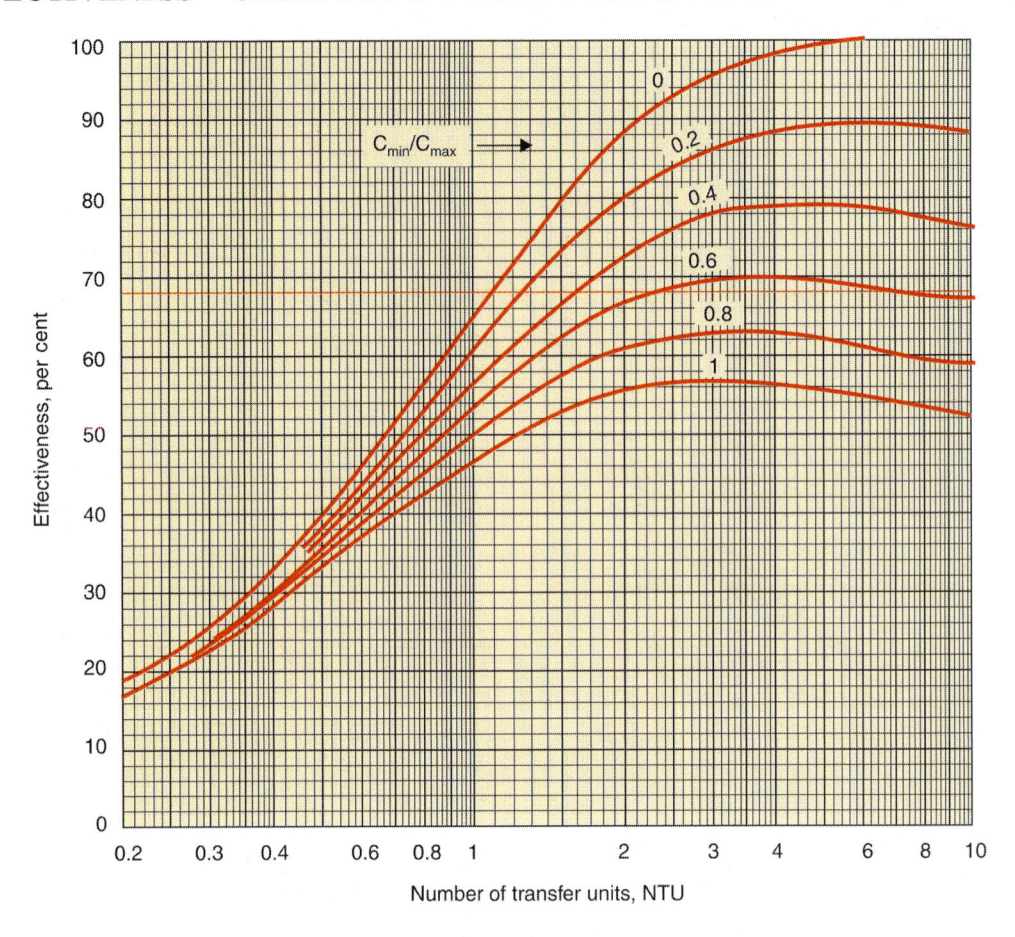

STORAGE TYPE HEAT EXCHANGER—TEMPERATURE VARIATION ALONG FLOW DIRECTION

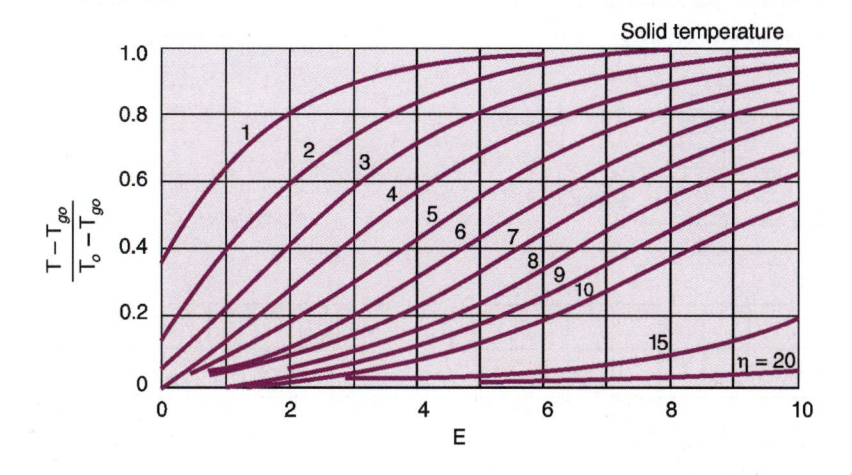

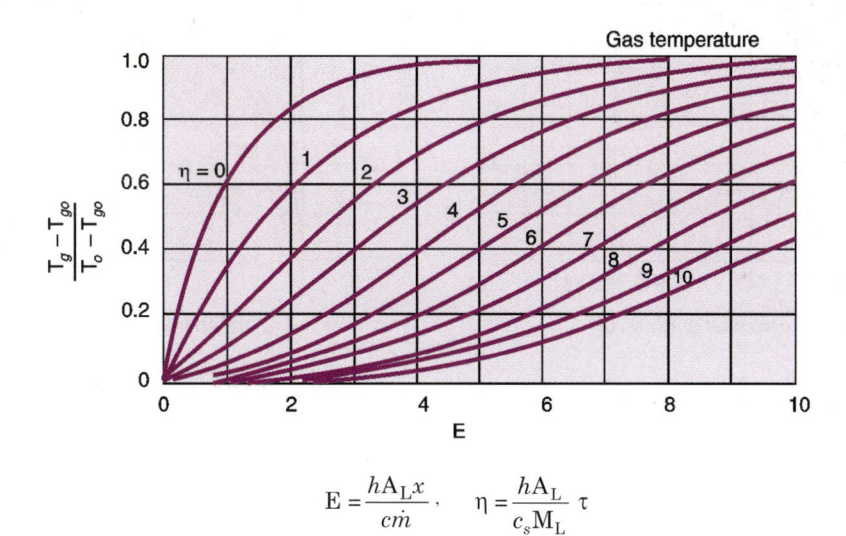

$$E = \frac{hA_L x}{c\dot{m}}, \quad \eta = \frac{hA_L}{c_s M_L}\tau$$

Refer page 154 and 155 for notations.

STORAGE TYPE EXCHANGER—EFFECTIVENESS

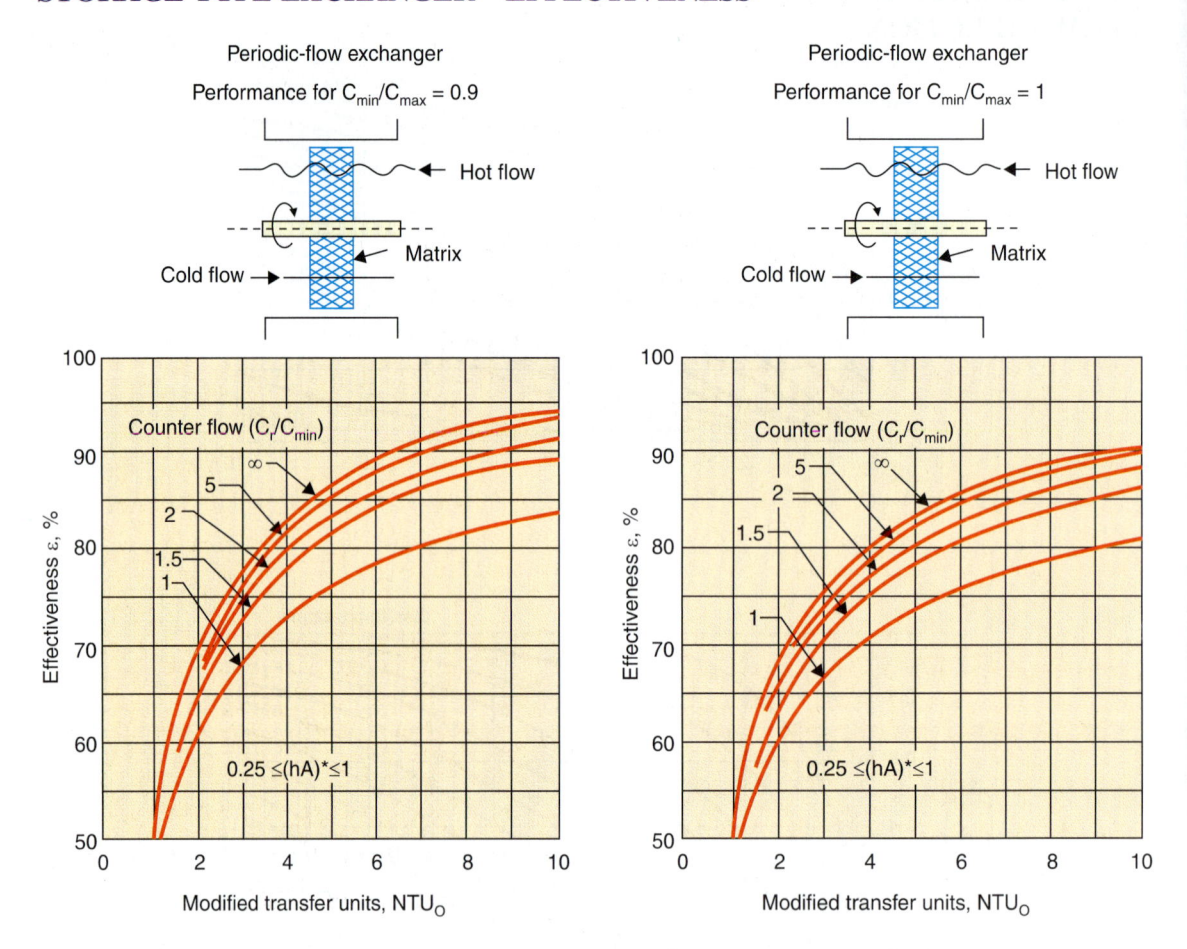

Periodic-flow exchanger
Performance for $C_{min}/C_{max} = 0.9$

Periodic-flow exchanger
Performance for $C_{min}/C_{max} = 1$

Note: For NTU_o and C_r refer to page 155.

HEAT AND MASS TRANSFER DATA BOOK

TRANSVERSE FIN HEAT EXCHANGER

Equations	Notations
The value of the y co-ordinate $\dfrac{h_o D_e}{k} \times Pr^{-0.333}$ can be read from the chart next page. To facilitate the use of computer and also for better accuracy curve fitted equation for the line on graph next page is (y) $$y = \exp\left[\ln 30 + 0.731 \ln\left(\dfrac{Re}{3000}\right)\right]$$ $$Re = \dfrac{G \cdot D_e}{\mu}$$ $$G = \dfrac{m}{NT \times TP \times L}$$ $$AF = \dfrac{\pi}{4} \times 2 \times NF \times [FD^2 - OD^2]$$ $$AB = \pi \times OD[L - NF \times t]$$ $$P = (2 \times 2 \times NF \times FL) + 2L$$ $$D_e = \dfrac{2[AF + AB]}{\pi \times P}$$ $$h_o = y \times Pr^{\frac{1}{3}} \times k / D_e$$ $$h_{oe} = \dfrac{h_o \times h_{fo}}{h_o + h_{fo}}$$ h_i is to be determined from pipe flow equations. $$h_{ie} = \dfrac{h_i \times h_{fi}}{h_i + h_{fi}}$$ $$h = \dfrac{\left[\{(\eta_f \times AF) + AB\} \times h_{oe}\right]}{\pi \times ID \times L}$$ $$U = \dfrac{h \times h_{ie}}{h + h_{ie}}$$ $$A = \dfrac{Q}{U(LMTD)_e}$$	Refer figure next page for FD, OD, ID, TP, FL, FP and t m — mass flow over tube bank, kg/s L — height of tube bank NT — number of tubes in bank G — mass flux, kg/sm^2 NF — Number of fins in length L, NF = L / FP D_e — Equivalent Diameter, m AF — Fin surface area, m^2 AB — Bare area over the fin leaving fin base, m^2 P — perimeter h_{fo} — Fouling factor on the outside. h_{fi} — Fouling factor on the inside. h_{oe} — effective convection coefficient on outside. h_{ie} — effective convection coefficient on inside h_i — convection coefficient inside tubes h — convection coefficient based on inside area η_f — fin efficiency, refer section on fins U — overall heat transfer coefficient Q — heat flow required calculated from mass flow and temperature change across the bank. A — Tube inside area required for the heat exchange Q

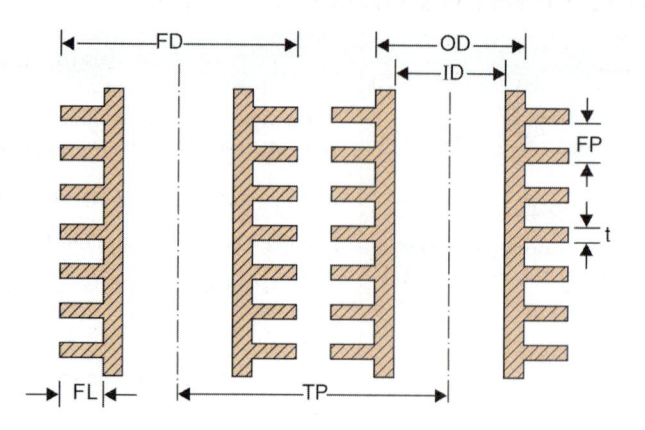

FIN CONFIGURATION

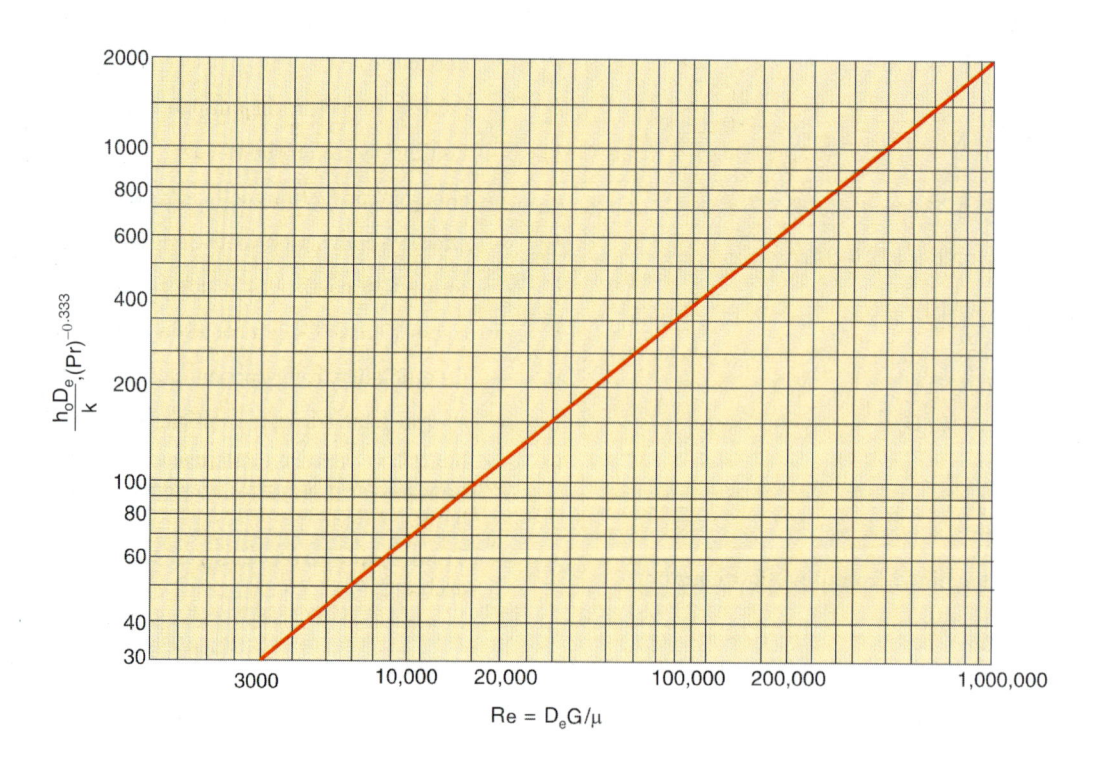

Transverse-fin heat transfer

NUMBER OF TUBES PER SHELL FOR DIFFERENT TUBE SHEET LAYOUT

	Square Pitch Arrangement								Triangular Pitch Arrangement							
Tube ID, OD, mm	19.0 25.4		25.4 31.0		31.0 39.7		38.1 47.6		19.0 25.4		25.4 31.0		31.0 39.7		38.1 47.6	
Shell Dia mm	1P	2P	1P	2P	1P	2P	1P	2P	1P	2P	1P	2P	1P	2P	1P	2P
203.2	32	26	21	16	—	—	—	—	37	30	21	16	—	—	—	—
254.0	52	52	32	32	16	12	—	—	61	52	32	32	20	18	—	—
304.8	81	76	48	45	30	24	16	16	92	82	55	52	32	30	18	34
336.6	97	90	61	56	32	30	22	22	109	106	68	66	38	36	27	22
387.4	137	124	81	76	44	40	29	29	151	138	91	86	54	51	36	34
438.2	177	166	112	112	56	53	39	39	203	196	131	118	69	66	48	44
489.0	224	220	138	132	78	73	50	48	262	250	163	152	95	91	61	58
539.8	277	270	177	166	96	90	62	60	316	302	199	188	117	112	76	72
590.6	341	324	213	208	127	112	78	74	384	376	241	232	140	136	95	91
635.0	413	394	260	252	140	135	94	90	470	452	294	282	170	164	115	110
685.8	481	460	300	288	166	160	112	108	559	534	349	334	202	196	136	131
736.6	553	526	341	326	193	188	131	127	630	604	397	376	235	228	160	154
787.4	657	640	406	398	226	220	151	146	745	728	472	454	275	270	184	177
838.2	749	718	465	460	258	252	176	170	856	830	538	522	315	305	215	206
889.0	845	824	522	518	293	287	202	196	970	938	608	592	357	348	246	238
939.8	934	914	596	574	334	322	224	220	1074	1044	674	664	407	390	275	268
990.6	1049	1024	665	644	370	362	252	246	1206	1176	766	736	449	436	307	299

P — number of tube passes

BASIC EQUATION FOR DIFFUSION COEFFICIENT

$$D_{AB} = \frac{0.04357\, T^{1.5}}{P[V_A^{0.333} + V_B^{0.333}]^2} \sqrt{\frac{1}{M_A} + \frac{1}{M_B}}$$

D_{AB} — Diffusion coefficient of material A into material B, m²/s
T — Temperature, K
P — Pressure in Pascal
V_A, V_B — Atomic volume of A and B – refer table
M_A, M_B — Molecular mass of A and B

Material	Atomic Volume
Air	29.9
Carbon	14.8
Carbon Dioxide	34.0
Hydrogen	14.3
Nitrogen (N)	15.6
Oxygen	7.40
Oxygen, in union with S, P, N	8.30
Oxygen, in acids	12.0
Water	18.8
Chlorine	21.6
Bromine	27.0
Iodine	37.0
Fluorine	8.70
Sulphur (S)	25.6
Phosphorus (P)	27.0

HEAT AND MASS TRANSFER DATA BOOK

MASS TRANSFER

Conditions	Correlation and Validity	Notations
MOLECULAR DIFFUSION 1. Fick's Law—steady state equimolal counter diffusion of a component A into another component B (Solids, liquids, gases)	$\dfrac{N_a}{A} = -D_{ab}\dfrac{\partial C_a}{\partial y}$ $\dfrac{\dot{m}_a}{A} = -D_{ab}\dfrac{\partial \rho_a}{\partial y}$	$\dfrac{N_a}{A}$ = mole flux of diffusing component A in kg mole/m² s. $\dfrac{\dot{m}_a}{A}$ — mass flux of diffusing component A in kg/m² s. A — Area, m², normal to the direction of diffusion – y.
2. Steady state equimolal counter diffusion between liquids along y direction $\dfrac{N_a}{A} = \dfrac{N_b}{A}$	$\dfrac{N_a}{A} = D_{ab}\dfrac{C_{a1}-C_{a2}}{y_2-y_1}$ $\dfrac{\dot{m}_a}{A} = D_{ab}\dfrac{\rho_{a1}-C_{a2}}{y_2-y_1}$	C_a — concentration of component A per unit volume of mixture of components A and B. kg mole/m³. When used with $\dfrac{N_a}{A}$ and kg/m³ when used with $\dot{m}/A$. D_{ab} — diffusion coefficient when component A diffuses into component B, m²/s. Refer p. 180 and 181.
3. Fick's law for **gases** (steady state equimolal counter diffusion) $N_a = N_b$	$\dfrac{\dot{m}_a}{A} = -D_{ab}\dfrac{1}{R_a T}\dfrac{dP_a}{dy}$	P_a, P_b partial pressures of components A and B in the gas mixture R_a — gas constant for component A
4. Equimolal counter diffusion of gases	$\dfrac{N_a}{A} = -D_{ab}\cdot\dfrac{1}{RT}\dfrac{dP_a}{dy}$ $\dfrac{\dot{m}_a}{A} = \dfrac{D_{ab}}{R_a T}\cdot\dfrac{P_{a1}-P_{a2}}{y_2-y_1}$	R — universal gas constant T — Temperature in K P — Total pressure = $P_a + P_b$ P_b — Partial pressure of non-diffusing component
5. Steady state diffusion of component A into a stagnant component B (gas/vapour) $N_b = 0$	$\dfrac{\dot{m}_a}{A} = D_{ab}\dfrac{P}{R_a T\cdot(y_2-y_1)}$ $\times \ln(P_{b2}/P_{b1})$ $= D_{ab}\cdot\dfrac{P}{R_a T\cdot(y_2-y_1)P_{bm}}$ Note: $P_{a1}-P_{a2} = P_{b2}-P_{b1}$	P_{bm} — logarithmic mean partial pressure difference $(P_{b2}-P_{b1})/\ln(P_{b2}/P_{b1})$ ρ — density
6. Steady state diffusion of component A into a stagnant mixture of components B, C, etc.	$\dfrac{\dot{m}_a}{A} = D_{a\,mix}\dfrac{P}{R_a T\cdot(y_2-y_1)}$ $\times \ln\dfrac{P_{mix2}}{P_{mix1}}$	$D_{a.mix}$ — diffusion coefficient when component A diffuses into a mixture of components B, C, D, etc.

(Contd.)

Conditions	Correlation and Validity	Notations
7. Steady state diffusion of liquid A into stagnant liquid B. ($N_b = 0$)	$\dfrac{N_a}{A} = D_{a\cdot mix} \cdot \dfrac{\rho_{a1} - \rho_{a2}}{y_2 - y_1} \cdot \dfrac{C}{C_{bm}}$	$D_{a.mix} = \dfrac{1}{\dfrac{N_b}{D_{ab}} + \dfrac{N_C}{D_{ac}} + \dfrac{N_D}{D_{ad}} + \ldots}$
8. Diffusion through solid boundary (similar to conduction)	Plate $\dfrac{m_a}{A} = \dfrac{D_{ab}}{L} \cdot (\rho_{a1} - \rho_{a2})$ Hollow cylinder of length l, $\dot{m}_a = \dfrac{2\pi D_{ab} \cdot l (\rho_{a1} - \rho_{a2})}{\ln \dfrac{r_2}{r_1}}$ Sphere $\dot{m}_a = \dfrac{4\pi r_1 r_2 (\rho_{a1} - \rho_{a2})}{(r_2 - r_1)}$	N_b, N_c, N_D mole fraction of component B, C, D in the mixture before diffussion. P — total pressure P_{mix} — sum of the partial pressure of components other than the diffusing component. $C = C_a + C_b$ $C_{bm} = (C_{b2} - C_{b1})/\ln(C_{b2}/C_{b1})$ $\dot{m}_a$ — mass diffusing rate ρ_{a1}, ρ_{a2} — concentration, kg/m³
Transient Diffusion into a semi infinite solid.	Note: The charts, tables and equations used for transient conduction can be used with the following equivalents. $\dfrac{T - T_\infty}{T_i - T_\infty} = \dfrac{\rho_a - \rho_{a\infty}}{\rho_{ai} - \rho_{a\infty}}$ $\dfrac{\alpha \tau}{L^2} = \dfrac{D_{ab} \tau}{L^2}$ $\dfrac{hL}{k} = \dfrac{h_m L}{D_{ab}}$ $\dfrac{x}{2\sqrt{\alpha \tau}} = \dfrac{x}{2\sqrt{D_{ab} \tau}}$	$\rho_{a\infty}$ — concentration of component A in the free stream ρ_{ai} — initial concentration of component A h_m — convective mass transfer coefficient m/s defined through, $\dfrac{\dot{m}}{A} = h_m [\rho_{a1} - \rho_{a2}]$
9. Convective mass transfer: 9.1 External flow: Flat Plate: Laminar Turbulent	Note: For situations where heat and mass transfer analogy is valid, the corresponding mass transfer correlations can be obtained by substituting Sh and Sc in the place of Nu and Pr. $\delta_{mx} = \delta_x \cdot Sc^{-0.333}$ $Sh_x = 0.332\, Re_x^{0.5}\, Sc^{0.333}$ $Sh = 0.664\, Re^{0.5}\, Sc^{0.333}$ $\dfrac{h_m}{u_\infty} Sc^{0.67} = \dfrac{0.322}{Re^{0.5}} = \dfrac{C_f}{2} = \dfrac{f}{8}$ $\delta_m \simeq \delta_t \simeq \delta_h$	D — diffusion coefficient, m²/s δ_{mx} — mass transfer boundary layer thickness at x δ_x — hydrodynamic boundary layer thickness at x Sh_x — local sherwood number, $h_m x/D$ Sc — Schmidt number, $\mu/\rho D$, refer p. 182, 183, 184 Re_x — local Reynolds number Sh — average Sherwood number h_m — convective mass transfer coefficient, m/s. St_m — mass transfer Stanton Number $= h_m/u_\infty$.

HEAT AND MASS TRANSFER DATA BOOK

Conditions	Correlation and Validity	Notations
Flat Plate Turbulent	$Sh_x = 0.0296 \, Re_x^{0.8} \cdot Sc^{0.333}$ $0.6 < Sc < 3000$ $Sh = 0.037 \, Re^{0.8}$ $5 \times 10^5 < Re < 10^8$ $0.6 < Sc < 3000$ Short laminar length	C_f — drag coefficient h_m — convective mass transfer coefficient, u_∞ — free stream velocity f — friction factor—equations for the factor are given in the convection chapter
9.2 Combined-Laminar Turbulent flow	$Sh = (0.037 \, Re^{0.8} - 871) \, Sc^{0.33}$ $\dfrac{h_m}{u_\infty} \cdot Sc^{0.67} = \dfrac{0.0286}{Re^{0.2}} = \dfrac{C_f}{2}$	h — heat transfer coefficient, W/m²K ρ — density, kg/m³ c_p — specific heat at constant pressure, J/kg K
9.3 Internal flow Turbulent	$Sh = 0.023 \, Re^{0.83} \, Sc^{0.44}$ $2000 < Re < 35000$, $0.6 < Sc < 2.5$ $\dfrac{h_m}{u_\infty} \cdot Sc^{0.67} = \dfrac{f}{8}$, Rough pipes	L_e — Lewis Number = Sc/Pr = α/D α — thermal diffusivity, m²/s h_{m1} — mass transfer coefficient from A to interface
10. Simultaneous convective heat and mass transfer	$\dfrac{h}{h_m} = c_p \, \rho \, Le^{0.67}$	h_{m2} — mass transfer coefficient from interface to medium B
11. Mass transfer through liquid gas interface	$\dfrac{1}{h_m} = \dfrac{1}{h_{m1} H_1} + \dfrac{1}{h_{m2}}$	H — Henry's constant (see tables) page 185
12. Packed beds	Refer packed beds under "CONVECTION".	

HUMIDIFICATION

Conditions	Equation	Notations	
1. Water and air in equilibrium—air at constant temperature	$P_w - P_\infty = \dfrac{h}{h_{fg} h_m}(T_\infty - T_w)$ $T_\infty - T_w = \dfrac{h_{fg}\left[\dfrac{P_w}{T_w} - \dfrac{P_\infty}{T_\infty}\right]}{R_w \rho c_p Le^{0.67}}$ $T_\infty - T_w \simeq \dfrac{h_{fg}[P_w - P_\infty]}{R_w \rho c_p . Le^{0.67} T_f}$ $T_\infty - T_w = \dfrac{0.622 h_{fg}}{c_p Le^{0.67}}\left[\dfrac{P_w}{P} - \dfrac{P_\infty}{P}\right]$ $\dfrac{Y_w - Y_\infty}{T_a - T_w} = \dfrac{c_p}{h_{fg}} . Le^{0.67}$	P_w P_∞ h h_{fg} Y_w Y_∞ p T_∞	partial pressure of water vapour at wet bulb temperature, T_w partial pressure of water vapour in air convective heat transfer coefft. for air to water enthalpy of evaporation for water at temperature, T_w abs. humidity of air at temp. T_w abs. humidity of air at temp. T_∞ total pressure temperature of air (dry bulb)
2. Water and air in equilibrium, water at constant temperature	$Y_a - Y_g = \dfrac{1}{h_{fg}}[c_p + Y_g c_{pw}][T_g - T_a]$	T_f R_w Y_a	—Mean temperature —gas constant of water vapour. abs. humidity of air at the final equilibrium air temperature T_a, kg/kg
3. Adiabatic humidification	$Z = \dfrac{G}{h_{mg}} \ln [(Y_a - Y_1)/(Y_a - Y_2)]$	Y_g c_p c_{pw} Z G h_{mg} a Y	abs. humidity of the initial air at the initial air temperature T_g specific heat of air specific heat of water vapour height of tower mass velocity of air per unit area kg/m²s gas phase mass transfer coefficient, kg/m²s interfacial area per unit volume of tower packing, 1/m abs. humidity of air

HUMIDIFICATION *(Contd.)*

Conditions	Equation	Notations	
4. Dehumidification	$h_l(T_i - T_l) = h_g(T_g - T_i)$ $+ h_{fg} h_{mg}(Y_g - Y_i)$	Y_1	humidity of air at entry to tower
		Y_2	humidity of air at exit of tower
		A	total area for mass transfer = a s Z
		$\dot{m}$	mass flux per unit time
		s	cross sectional area of tower
		a	interfacial area unit volume
		h_l	liquid phase heat transfer coefficient, $W/m^2 K$
		h_g	gas phase heat transfer coefficient, $W/m^2 K$
		T_i	temperature at the interface
		T_g	bulk gas temperature (dry bulb)
		T_l	liquid layer temperature
		Y_i	abs. humidity at temperature T_i.

DIFFUSION COEFFICIENTS

Diffusing material, Solute	Medium of Diffusion, Solvent	Temperature °C	Concentration of Solute, kg mole/m^3	Diffusion Coefficient, D, m^2/s
Acetic acid	Water	12.5	0.01	0.9111×10^{-9}
Acetic acid	Water	12.5	1.0	0.8222×10^{-9}
Acetic acid	Water	18.0	1	0.9611×10^{-9}
Ammonia	Water	5	3.5	1.2389×10^{-9}
Ammonia	Water	15	1	1.7694×10^{-9}
Carbon dioxide	Water	10	$\simeq 0$	1.4611×10^{-9}
Carbon dioxide	Water	20	$\simeq 0$	1.7694×10^{-9}
Carbon dioxide	Ethanol	17	$\simeq 0$	3.1944×10^{-9}
Chlorine	Water	16	0.12	1.2611×10^{-9}
Chloroform	Ethanol	20	2	1.2500×10^{-9}
Copper	Aluminium	462	—	0.8472×10^{-9}
Ethanol	Water	10	0.05	0.8306×10^{-9}
Ethanol	Water	16	2	0.9000×10^{-9}
Ethanol	Water	10	3.75	0.5000×10^{-9}
Hydro chloric acid	Water	16	0.5	2.4416×10^{-9}
Hydro chloric acid	Water	0	2	1.8000×10^{-9}
Hydro chloric acid	Water	10	2.5	2.5000×10^{-9}
Hydro chloric acid	Water	0	9	2.7028×10^{-9}
Hydro chloric acid	Water	10	9	3.3056×10^{-9}
Mercury	Lead	177	—	0.1956×10^{-9}
Mercury	Lead	197	—	0.5000×10^{-9}
Methanol	Water	15	$\simeq 0$	1.2806×10^{-9}
n-Butanol	Water	15	$\simeq 0$	0.7722×10^{-9}
Sodium Chloride	Water	18	0.05	1.2611×10^{-9}
Sodium Chloride	Water	18	0.2	1.2083×10^{-9}
Sodium Chloride	Water	18	1	1.2389×10^{-9}
Sodium Chloride	Water	18	3	1.3611×10^{-9}
Sodium Chloride	Water	18	5.4	1.5417×10^{-9}

DIFFUSION COEFFICIENTS AND SCHMIDT NUMBER

Diffusing Component Solute	Diffusion Medium Solvent	Temperature °C	Diffusion Coefficient D, m²/s	Schmidt Number Sc
Air	Ethanol	0	10.19×10^{-6}	—
Air	n-Butanol	26	8.69×10^{-6}	—
Air	n-Butanol	59	10.19×10^{-6}	—
Air	Ethyl acetate	26	8.69×10^{-6}	—
Air	Ethyl acetate	59	10.39×10^{-6}	—
Air	Aniline	26	8.69×10^{-6}	—
Air	Aniline	59	10.61×10^{-6}	—
Air	Chlorobenzene	26	7.39×10^{-6}	—
Air	Chlorobenzene	59	9.00×10^{-6}	—
Air	Toluene	26	7.39×10^{-6}	—
Air	Toluene	59	9.00×10^{-6}	—
Ammonia	Air	0	21.60×10^{-6}	0.634
Benzene	Air	0	7.50×10^{-6}	1.83
Benzene	Carbhon dioxide	0	5.14×10^{-6}	1.37
Benzene	Hydrogen	0	29.44×10^{-6}	3.26
Carbon dioxide	Air	0	11.89×10^{-6}	1.14
Carbon dioxide	Hydrogen	18	60.56×10^{-6}	0.158
Carbon dioxide	Oxygen	0	18.47×10^{-6}	—
Carbon disulphide	Air	20	8.81×10^{-6}	1.68
Carbon monoxide	Oxgyen	0	18.50×10^{-6}	—
Ether	Air	20	7.69×10^{-6}	1.93
Ethyl alcohol	Air	0	10.11×10^{-6}	1.36
Ethyl alcohol	Air	40.3	11.81×10^{-6}	1.45
Hydrogen	Air	0	54.72×10^{-6}	0.25
Hydrogen	Oxygen	14	77.50×10^{-6}	0.182
Hydrogen	Nitrogen	12.5	73.89×10^{-6}	0.187
Hydrogen	Methane	0	62.50×10^{-6}	—
Mercury	Nitrogen	19	3250×10^{-6}	0.00424
Oxygen	Air	0	15.28×10^{-6}	0.895
Oxygen	Nitrogen	12	20.17×10^{-6}	0.621
Water	Air	8	20.58×10^{-6}	0.615
Water	Air	16	28.06×10^{-6}	0.488
Water	Air	26	25.83×10^{-6}	—
Water	Air	59	30.56×10^{-6}	—

BINARY DIFFUSION COEFFICIENT AT ONE ATMOSPHERE

Material A	Material B	Temperature °C	Diffusion Coefficient, D_{ab}	Solubility, $kmol/m^3\ bar$
Aluminium	Copper	20	0.13×10^{-33}	
Acetone	Air	0	11×10^{-6}	
Air	Nitrogen	0	10×10^{-6}	
Cadmium	Copper	20	0.27×10^{-18}	
Carbon Dioxide	Nitrogen	20	16×10^{-6}	
Carbon Dioxide	Rubber	25	0.11×10^{-9}	40.15×10^{-3}
Helium	Silicon Dioxide	20	4×10^{-12}	0.45×10^{-3}
Hydrogen	Nickel	85	—	9.01×10^{-3}
Hydrogen	Iron	20	0.26×10^{-12}	
Naphthalene	Air	27	6.2×10^{-6}	
Nitrogen	Rubber	25	0.15×10^{-9}	1.56×10^{-3}
Oxygen	Rubber	25	0.21×10^{-9}	3.12×10^{-3}

Dilute Solutions

Material A	Material B	Temperature °C	Diffusion Coefficient, D_{ab}	Solubility, $kmol/m^3\ bar$
Acetone	Water	25	1.3×10^{-9}	
Caffeine	Water	25	0.63×10^{-9}	
Carbon Dioxide	Water	25	2.0×10^{-9}	
Ethanol	Water	25	0.12×10^{-8}	
Glucose	Water	25	0.69×10^{-9}	
Glycerol	Water	25	0.94×10^{-9}	
Hydrogen	Water	25	6.3×10^{-9}	
Nitrogen	Water	25	2.6×10^{-9}	
Oxygen	Water	25	2.4×10^{-9}	

SCHMIDT NUMBER OF VARIOUS SUBSTANCES AT 20°C, Sc*

when in dilute solution in water, ethyl alcohol or benzene.†

Solute	Solvent	Sc*
Acetic acid	Water	1140
Acetic acid	Benzene	384
Acetylene	Water	645
Allyl alcohol	Water	1080
Ammonia	,,	570
Bromine	,,	840
Butyl alcohol	,,	1310
Carbon dioxide	,,	559
–do–	Ethyl alcohol	445
Chlorine	Water	824
Chloroform	Ethyl alcohol	1230
Chloroform	Benzene	350
Ethylene dichlorde	Benzene	301
Ethyl alcohol	Water	1005
Glycerol	,,	1400
Hydrogen	,,	196
Hydrogen Sulphide	,,	712
Hydrochloric acid	,,	381
Hydroquinone	,,	1300
Lactose	,,	2340
Manitol	,,	1730
Methyl alcohol	,,	785
Nitrogen	,,	613
Nitric acid	,,	390
Nitrous oxide	,,	665
Oxygen	,,	558
Phenol	,,	1200
Phenol	Ethyl acohol	1900
Phenol	Benzene	479
Propyl alcohol	Water	1150
Pyrogallol	,,	1440
Raffinose	,,	2720
Resorcinol	,,	1260
Sodium chloride	,,	745
Sodium hydroxide	,,	665
Sucrose	,,	2230
Sulphuric acid	,,	580
Urea	,,	944
Urethane	,,	1090

† Schmidt Number at any other temperature, may be found from the following relationship:

$$\frac{Sc}{Sc^*} = \left(\frac{\mu}{\mu^*}\right)^2 \left(\frac{\rho^*}{\rho}\right)\left(\frac{T^*}{T}\right)$$

where
- μ, absolute viscosity at absolute temperature T
- ρ, density at absolute temperature T
- ρ^*, density at 20°C
- μ^*, absolute viscosity at 20°C
- $T^* = 293$ K

SCHMIDT NUMBER OF GASES IN DILUTE MIXTURE WITH AIR†

Gas	Molecular Weight M	Schmidt Number Sc^*
Acetic acid	60.05	1.24
Acetone	58.08	1.60
Ammonia	17.03	0.61
Benzene	78.11	1.71
Bromobenzol	157.02	1.97
Butane	58.12	1.77
Carbon dioxide	44.01	0.96
Carbon disulphide	76.13	1.48
Carbon tetrachloride	153.84	2.13
Chlorine	70.90	1.42
Chlorobenzene	112.56	2.13
Chloropicrin	164.39	2.13
Ethane	30.07	1.22
Ethyl acetate	88.10	1.84
Ethyl alcohol	46.07	1.30
Ethyl ether	74.12	1.70
Ethylene bromide	187.88	1.97
Hydrogen	2.016	0.22
Methane	16.04	0.84
Methyl alcohol	32.04	1.00
Methyl acetate	74.08	1.57
Napthalene	128.16	2.57
Nitrogen	28.02	0.98
n-Butyl alcohol	74.12	1.88
n-Octane	114.22	2.62
n-Propyl acetate	102.13	1.97
n-Propyl alcohol	60.09	1.55
Oxygen	32.00	0.74
Pentane	72.15	1.97

(Contd.)

HEAT AND MASS TRANSFER DATA BOOK

Gas	Molecular Weight M	Schmidt Number Sc*
Phosgene	98.92	1.65
Propane	44.09	1.51
Steam	18.016	0.60
Sulphur dioxide	64.06	1.28
Toluene	92.13	1.86

† Schmidt Numbers of gases in dilute mixture with air may be deduced from the following approximate relationship :

$$Sc^* \simeq 0.145\, M^{0.556}$$

Schmidt number of a gas mixed with air in any proportion may be determined in the following way :

$$Sc = \frac{v}{v^*}\, Sc^*$$

where
  v, kinematic viscosity of the mixture, m²/s
  v^*, kinematic viscosity of the component in a dilute mixture at the mixture temperature, m²/s.

HENRY'S CONSTANT FOR SELECTED GASES IN WATER AT MODERATE PRESSURE

$$H = P_A/C_A,\ C\text{---concentration}$$

T(K)	NH_3	Cl_2	H_2S	SO_2	CO_2	CH_4	O_2	H_2
273	21	265	260	165	710	22,880	25,500	58,000
280	23	365	335	210	960	27,800	30,500	61,500
290	26	480	450	315	1300	35,200	37,600	66,500
300	30	615	570	440	1730	42,800	45,700	71,600
310	—	755	700	600	2175	50,000	52,500	76,000
320	—	860	835	800	2650	56,300	56,800	78,000
323	—	890	870	850	2870	58,000	58,000	79,000

Collision function for diffusion

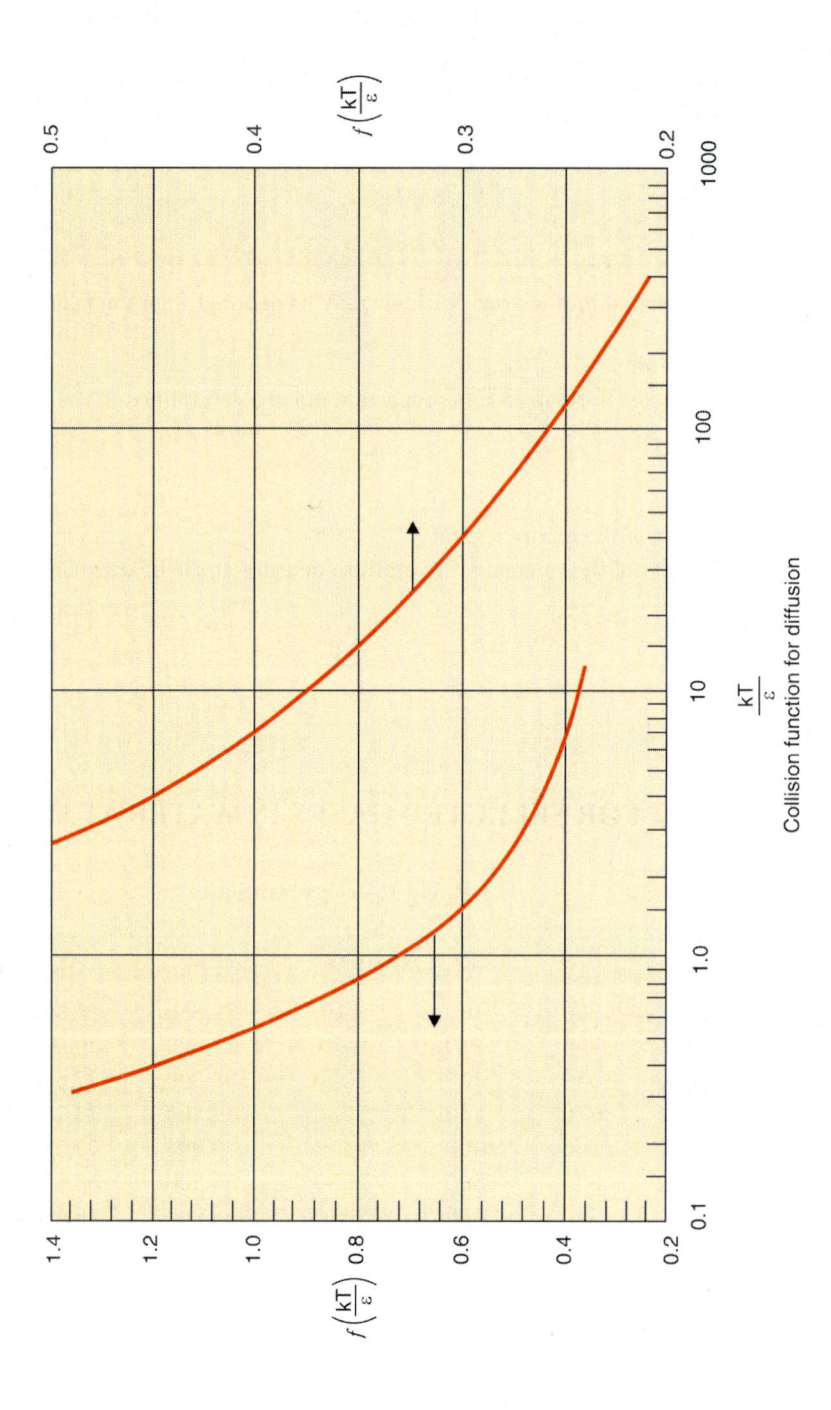

HEAT AND MASS TRANSFER DATA BOOK

Table 1: HP1 Properties of Heat Pipe Fluids – Lithium, Potassium, Sodium and Mercury

Lithium

T, °C	h_{fg}, kJ/kg	ρ_l kg/m^3	ρ_v kg/m^3	k_l W/m °C	μ_l cP	μ_v cP×10^2	P_v bar	c_{pv} kJ/kg °C	σ_l N/m×10^2
1030	20500	450	0.005	67	0.24	1.67	0.07	0.532	2.90
1130	20100	440	0.013	69	0.24	1.74	0.17	0.532	2.85
1230	20000	430	0.028	70	0.23	1.83	0.45	0.532	2.75
1330	19700	420	0.057	69	0.23	1.91	0.96	0.532	2.60
1430	19200	410	0.108	68	0.23	2.00	1.85	0.532	2.40
1530	18900	405	0.193	65	0.23	2.10	3.30	0.532	2.25
1630	18500	400	0.340	62	0.23	2.17	5.30	0.532	2.10
1730	18200	398	0.490	59	0.23	2.26	8.90	0.532	2.05

Potassium

350	2093	763.1	0.002	51.08	0.21	0.15	0.01	5.32	9.50
400	2078	748.1	0.006	49.08	0.19	0.16	0.01	5.32	9.04
450	2060	735.4	0.015	47.08	0.18	0.16	0.02	5.32	8.69
500	2040	725.4	0.031	45.08	0.17	0.17	0.05	5.32	8.44
550	2020	715.4	0.062	43.31	0.15	0.17	0.10	5.32	8.16
600	2000	705.4	0.111	41.81	0.14	0.18	0.19	5.32	7.86
650	1980	695.4	0.193	40.08	0.13	0.19	0.35	5.32	7.51
700	1969	685.4	0.314	38.08	0.12	0.19	0.61	5.32	7.12
750	1938	675.4	0.486	36.31	0.12	0.20	0.99	5.32	6.72
800	1913	665.4	0.716	34.81	0.11	0.20	1.55	5.32	6.32
850	1883	653.1	1.054	33.31	0.10	0.21	2.34	5.32	5.92

Sodium

500	4370	828.1	0.003	70.08	0.24	0.18	0.01	9.04	1.51
600	4243	805.4	0.013	64.52	0.21	0.19	0.04	9.04	1.42
700	4090	763.5	0.050	60.81	0.19	0.20	0.15	9.04	1.33
800	3977	757.3	0.134	57.81	0.18	0.22	0.47	9.04	1.23
900	3913	745.4	0.306	53.35	0.17	0.23	1.25	9.04	1.13
1000	3827	725.4	0.667	49.08	0.16	0.24	2.81	9.04	1.04
1100	3690	690.8	1.306	45.08	0.16	0.25	5.49	9.04	0.95
1200	3527	669.0	2.303	41.08	0.15	0.26	9.59	9.04	0.86
1300	3477	654.0	3.622	37.08	0.15	0.27	15.91	9.04	0.77

Mercury

150	308.8	13280	0.01	9.99	1.09	0.39	0.01	1.04	4.45
250	303.8	12995	0.60	11.23	0.96	0.48	0.18	1.04	4.15
300	301.8	12880	1.73	11.73	0.93	0.53	0.44	1.04	4.00
350	298.9	12763	4.45	12.18	0.89	0.61	1.16	1.04	3.82
400	296.3	12656	8.75	12.58	0.86	0.66	2.42	1.04	3.74
450	293.8	12508	16.80	12.96	0.83	0.70	4.92	1.04	3.61
500	291.8	12308	28.60	13.31	0.80	0.75	8.86	1.04	3.41
550	288.8	12154	44.92	13.62	0.79	0.81	15.03	1.04	3.25
600	286.3	12054	65.75	13.87	0.78	0.87	23.97	1.04	3.15
650	283.5	11962	94.29	14.15	0.78	0.95	34.95	1.04	3.03
750	277.0	11800	170.00	14.80	0.77	1.10	63.00	1.04	2.75

h_{fg} – Latent heat, ρ_l – Density of liquid, ρ_v – Density of vapour,

k_l – Thermal conductivity of liquid, μ_l – Viscosity of liquid,

μ_v – Viscosity of vapour, P_v – Vapour pressure,

c_{pv} – Specific heat of vapour, σ_l – Surface tension of liquid

Table 2: HP2 Properties of Heat Pipe Fluids – Caesium, Flutec PP9, High Temperature Organic (Diphenyl – Diphenyl Oxideutectic) and Flutec PP2

Caesium

T, °C	h_{fg}, kJ/kg	ρ_l kg/m³	ρ_v kg/m³	k_l W/m °C	μ_l cP	μ_v cP×10²	P_v bar	c_{pv} kJ/kg °C	σ_l N/m×10²
375	530.4	1740	0.01	20.76	0.25	2.20	0.02	1.56	5.81
425	520.4	1730	0.01	20.51	0.23	2.30	0.04	1.56	5.61
475	515.2	1720	0.02	20.02	0.22	2.40	0.09	1.56	5.36
525	510.2	1710	0.03	19.52	0.20	2.50	0.16	1.56	5.11
575	502.8	1700	0.07	18.83	0.19	2.55	0.36	1.56	4.81
625	495.3	1690	0.10	18.18	0.18	2.60	0.57	1.56	4.51
675	490.2	1680	0.18	17.48	0.17	2.67	1.04	1.56	4.21
725	485.2	1670	0.26	16.83	0.17	2.75	1.52	1.56	3.91
775	477.8	1655	0.40	16.18	0.16	2.88	2.46	1.56	3.66
825	470.3	1640	0.55	15.53	0.16	2.90	3.41	1.56	3.41

Flutec PP9

T, °C	h_{fg}, kJ/kg	ρ_l kg/m³	ρ_v kg/m³	k_l W/m °C	μ_l cP	μ_v cP×10²	P_v bar	c_{pv} kJ/kg °C	σ_l N/m×10²
−30	103.0	2098	0.01	0.060	5.77	0.82	0.00	0.80	2.36
0	98.4	2029	0.01	0.059	3.31	0.90	0.00	0.87	2.08
30	94.5	1960	0.12	0.057	1.48	1.06	0.01	0.94	1.80
60	90.2	1891	0.61	0.056	0.94	1.18	0.03	1.02	1.52
90	86.1	1822	1.93	0.054	0.65	1.21	0.12	1.09	1.24
120	83.0	1753	4.52	0.053	0.49	1.23	0.28	1.15	0.95
150	77.4	1685	11.81	0.052	0.38	1.26	0.61	1.23	0.67
180	70.8	1604	25.13	0.051	0.30	1.33	1.58	1.30	0.40
225	59.4	1455	63.27	0.0049	0.21	1.44	4.21	1.41	0.01

High Temperature Organic (Diphenyl – Diphenyl Oxideutectic)

T, °C	h_{fg}, kJ/kg	ρ_l kg/m³	ρ_v kg/m³	k_l W/m °C	μ_l cP	μ_v cP×10²	P_v bar	c_{pv} kJ/kg °C	σ_l N/m×10²
100	354.0	992	0.03	0.131	0.97	0.67	0.01	1.34	3.50
150	338.0	951	0.22	0.125	0.57	0.78	0.05	1.51	3.00
200	321.0	905	0.94	0.119	0.39	0.89	0.25	1.67	2.50
250	301.0	858	3.60	0.113	0.27	1.00	0.88	1.81	2.00
300	278.0	809	8.74	0.106	0.20	1.12	2.43	1.95	1.50
350	251.0	755	19.37	0.099	0.15	1.23	5.55	2.03	1.00
400	219.0	691	41.89	0.093	0.12	1.34	10.90	2.11	0.50
450	185.0	625	81.00	0.086	0.10	1.45	19.00	2.19	0.03

Flutec PP2

T, °C	h_{fg}, kJ/kg	ρ_l kg/m³	ρ_v kg/m³	k_l W/m °C	μ_l cP	μ_v cP×10²	P_v bar	c_{pv} kJ/kg °C	σ_l N/m×10²
−30	106.2	1942	0.13	0.637	5.200	0.98	0.01	0.72	1.90
−10	103.1	1886	0.44	0.626	3.500	1.03	0.02	0.81	1.71
10	99.8	1829	1.39	0.613	2.140	1.07	0.09	0.92	1.52
30	96.3	1773	2.96	0.601	1.435	1.12	0.22	1.01	1.32
50	91.8	1716	6.43	0.588	1.005	1.17	0.39	1.07	1.13
70	87.0	1660	11.79	0.575	0.720	1.22	0.62	1.11	0.93
90	82.1	1599	21.99	0.563	0.543	1.26	1.43	1.17	0.73
110	76.5	1558	34.92	0.550	0.429	1.31	2.82	1.25	0.52
130	70.3	1515	57.21	0.537	0.314	1.36	4.83	1.33	0.32
160	59.1	1440	103.63	0.518	0.167	1.43	8.76	1.45	0.01

h_{fg} – Latent heat, ρ_l – Density of liquid, ρ_v – Density of vapour,

k_l – Thermal conductivity of liquid, μ_l – Viscosity of liquid,

μ_v – Viscosity of vapour, P_v – Vapour pressure,

c_{pv} – Specific heat of vapour, σ_l – Surface tension of liquid

Table 3: HP3 Properties of Heat Pipe Fluids – Heptane, Acetone, Methanol and Ammonia

Heptane

T, °C	h_{fg}, kJ/kg	ρ_l kg/m^3	ρ_v kg/m^3	k_l W/m °C	μ_l cP	μ_v cP×10^2	P_v bar	c_{pv} kJ/kg °C	σ_l N/m×10^2
−20	384.0	715.5	0.01	0.143	0.69	0.57	0.01	0.83	2.42
0	372.6	699.0	0.17	0.141	0.53	0.60	0.02	0.87	2.21
20	362.2	683.0	0.49	0.140	0.43	0.63	0.08	0.92	2.01
40	351.8	667.0	0.97	0.139	0.34	0.66	0.20	0.97	1.81
60	341.5	649.0	1.45	0.137	0.29	0.70	0.32	1.02	1.62
80	331.2	631.0	2.31	0.135	0.24	0.74	0.62	1.05	1.43
100	319.6	612.0	3.71	0.133	0.21	0.77	1.10	1.09	1.28
120	305.0	592.0	6.08	0.132	0.18	0.82	1.85	1.16	1.10

Acetone

T, °C	h_{fg}, kJ/kg	ρ_l kg/m^3	ρ_v kg/m^3	k_l W/m °C	μ_l cP	μ_v cP×10^2	P_v bar	c_{pv} kJ/kg °C	σ_l N/m×10^2
−40	660.0	860.0	0.03	0.200	0.800	0.68	0.01	2.00	3.10
−20	615.6	845.0	0.10	0.189	0.500	0.73	0.03	2.06	2.76
0	564.0	812.0	0.26	0.183	0.395	0.78	0.10	2.11	2.62
20	552.0	790.0	0.64	0.181	0.323	0.82	0.27	2.16	2.37
40	536.0	768.0	1.05	0.175	0.269	0.86	0.60	2.22	2.12
60	517.0	774.0	2.37	0.168	0.226	0.90	1.15	2.28	1.86
80	495.0	719.0	4.30	0.160	0.192	0.95	2.15	2.34	1.62
100	472.0	689.6	6.94	0.148	0.170	0.98	40.43	2.39	1.34
120	426.1	660.3	11.02	0.135	0.148	0.99	6.70	2.45	1.07
140	394.4	631.8	18.61	0.126	0.132	1.00	10.49	2.50	0.81

Methanol

T, °C	h_{fg}, kJ/kg	ρ_l kg/m^3	ρ_v kg/m^3	k_l W/m °C	μ_l cP	μ_v cP×10^2	P_v bar	c_{pv} kJ/kg °C	σ_l N/m×10^2
−50	1194	843.5	0.01	0.210	1.700	0.72	0.01	1.20	3.26
−30	1187	833.5	0.01	0.208	1.300	0.78	0.02	1.27	2.95
−10	1182	818.7	0.04	0.206	0.945	0.85	0.04	1.34	2.63
10	1175	800.5	0.12	0.204	0.701	0.91	0.10	1.40	2.36
30	1155	782.0	0.31	0.203	0.521	0.98	0.25	1.47	2.18
50	1125	764.1	0.77	0.202	0.399	1.04	0.55	1.54	2.01
70	1085	746.2	1.47	0.201	0.314	1.11	1.31	1.61	1.85
90	1035	724.4	3.01	0.199	0.259	1.19	2.69	1.79	1.66
110	980	703.6	5.64	0.197	0.211	1.26	4.98	1.92	1.46
130	920	685.2	9.81	0.195	0.166	1.31	7.86	1.92	1.25
150	850	653.2	15.90	0.193	0.138	1.38	8.94	1.92	1.04

Ammonia

T, °C	h_{fg}, kJ/kg	ρ_l kg/m^3	ρ_v kg/m^3	k_l W/m °C	μ_l cP	μ_v cP×10^2	P_v bar	c_{pv} kJ/kg °C	σ_l N/m×10^2
−60	1343	714.4	0.03	0.294	0.36	0.72	0.27	2.050	4.062
−40	1384	690.4	0.05	0.303	0.29	0.79	0.76	2.075	3.574
−20	1338	665.5	1.62	0.304	0.26	0.85	1.93	2.100	3.090
0	1263	638.6	3.48	0.298	0.25	0.92	4.24	2.125	2.480
20	1187	610.3	6.69	0.286	0.22	1.01	8.46	2.150	2.133
40	1101	579.5	12.00	0.272	0.20	1.16	15.34	2.160	1.833
60	1026	545.2	20.49	0.255	0.17	1.27	29.80	2.180	1.367
80	891	505.7	34.13	0.235	0.15	1.40	40.90	2.210	0.767
100	699	455.1	54.92	0.212	0.11	1.60	63.12	2.260	0.500
120	428	374.4	113.16	0.184	0.07	1.89	90.44	2.292	0.150
−271	22.8	148.3	26.0	1.81	3.90	0.20	0.06	2.045	0.26
−270	23.6	140.7	17.0	2.24	3.70	0.30	0.32	2.699	0.19
−269	20.9	128.0	10.0	2.77	2.90	0.60	1.00	4.619	0.09
−268	4.0	113.8	8.5	3.50	1.34	0.90	2.29	6.642	0.01

h_{fg} – Latent heat, ρ_l – Density of liquid, ρ_v – Density of vapour,

k_l – Thermal conductivity of liquid, μ_l – Viscosity of liquid,

μ_v – Viscosity of vapour, P_v – Vapour pressure,

c_{pv} – Specific heat of vapour, σ_l – Surface tension of liquid

Table 4: HP4 Properties of Heat Pipe Fluids – Nitrogen, Helium, Ethanol, Water and Pentane

Nitrogen

T, °C	h_{fg}, kJ/kg	ρ_l kg/m³	ρ_v kg/m³	k_l W/m °C	μ_l cP	μ_v cP×10²	P_v bar	c_{pv} kJ/kg °C	σ_l N/m×10²
−203	210.0	830.0	1.84	0.150	2.48	0.48	0.48	1.083	1.054
−200	205.5	818.0	3.81	0.146	1.94	0.51	0.74	1.082	0.985
−195	198.0	798.0	7.10	0.139	1.51	0.56	1.62	1.079	0.870
−190	190.5	778.0	10.39	0.132	1.26	0.60	3.31	1.077	0.766
−185	183.0	758.0	13.68	0.125	1.08	0.65	4.99	1.074	0.662
−180	173.7	732.0	22.05	0.117	0.95	0.71	6.69	1.072	0.561
−175	163.2	702.0	33.80	0.110	0.86	0.77	8.37	1.070	0.464
−170	152.7	672.0	45.55	0.103	0.80	0.83	11.07	1.068	0.367
−160	124.4	603.0	80.90	0.089	0.72	1.00	19.37	1.063	0.185
−150	66.8	474.0	194.00	0.075	0.65	1.50	28.80	1.059	0.110

Helium

−271	22.8	148.3	26.0	1.81	3.90	0.20	0.06	2.045	0.26
−270	23.6	140.7	17.0	2.24	3.70	0.30	0.32	2.699	0.19
−269	20.9	128.0	10.0	2.77	2.90	0.60	1.00	4.619	0.09
−268	4.0	113.8	8.5	3.50	1.34	0.90	2.29	6.642	0.01

Ethanol

−30	939.4	825.0	0.02	0.177	3.40	0.75	0.01	1.25	2.76
−10	928.7	813.0	0.03	0.173	2.20	0.80	0.02	1.31	2.66
10	904.8	798.0	0.05	0.170	1.50	0.85	0.03	1.37	2.57
30	888.6	781.0	0.38	0.168	1.02	0.91	0.10	1.44	2.44
50	872.3	762.2	0.72	0.166	0.72	0.97	0.29	1.51	2.31
70	858.3	743.1	1.32	0.165	0.51	1.02	0.76	1.58	2.17
90	832.1	725.3	2.59	0.163	0.37	1.07	1.43	1.65	2.04
110	786.6	706.1	5.17	0.160	0.28	1.13	2.66	1.72	1.89
130	734.4	678.7	9.25	0.159	0.21	1.18	4.30	1.78	1.75

Water

20	2448	998.2	0.02	0.603	1.00	0.96	0.02	1.81	7.28
40	2402	992.3	0.05	0.630	0.65	1.04	0.07	1.89	6.96
60	2359	983.0	0.13	0.649	0.47	1.12	0.20	1.91	6.62
80	2309	972.0	0.29	0.668	0.36	1.19	0.47	1.95	6.26
100	2258	958.0	0.60	0.680	0.28	1.27	1.01	2.01	5.89
120	2200	945.0	1.12	0.682	0.23	1.34	2.02	2.09	5.50
140	2139	928.0	1.99	0.683	0.20	1.41	3.90	2.21	5.06
160	2074	909.0	3.27	0.679	0.17	1.49	6.44	2.38	4.66
180	2003	888.0	5.16	0.669	0.15	1.57	10.04	2.62	4.29
200	1967	865.0	7.87	0.659	0.14	1.65	16.19	2.91	3.89

Pentane

−20	390.0	663.0	0.01	0.149	0.344	0.51	0.10	0.825	2.01
0	378.3	644.0	0.75	0.143	0.283	0.53	0.24	0.874	1.79
20	366.9	625.5	2.20	0.138	0.242	0.58	0.76	0.922	1.58
40	355.5	607.0	4.35	0.133	0.200	0.63	1.52	0.971	1.37
60	342.3	585.0	6.51	0.128	0.174	0.69	2.28	1.021	1.17
80	329.1	563.0	10.61	0.127	0.147	0.74	3.89	1.050	0.97
100	295.7	537.6	16.54	0.124	0.128	0.81	7.19	1.088	0.83
120	269.7	509.4	25.20	0.122	0.120	0.90	13.81	1.164	0.68

h_{fg} – Latent heat, ρ_l – Density of liquid, ρ_v – Density of vapour,
k_l – Thermal conductivity of liquid, μ_l – Viscosity of liquid,
μ_v – Viscosity of vapour, P_v – Vapour pressure,
c_{pv} – Specific heat of vapour, σ_l – Surface tension of liquid

Table 5: Variation of Thermal Conductivity and Viscosity of R12, Dichlorodifluoromethane and R22, Chlorodifluoromethane

Temperature °C	R12, Dichlorodifluoromethane Viscosity, $\mu \times 10^6$ Liquid	R12, Dichlorodifluoromethane Viscosity, $\mu \times 10^6$ Vapour	R12, Dichlorodifluoromethane Conductivity, $k \times 10^3$ Liquid	R12, Dichlorodifluoromethane Conductivity, $k \times 10^3$ Vapour	R22, Chlorodifluoromethane Viscosity, $\mu \times 10^6$ Liquid	R22, Chlorodifluoromethane Viscosity, $\mu \times 10^6$ Vapour	R22, Chlorodifluoromethane Conductivity, $k \times 10^3$ Liquid	R22, Chlorodifluoromethane Conductivity, $k \times 10^3$ Vapour
−70	584.0	7.97	103.0	5.50	507.6	8.52	127.6	5.68
−60	505.1	8.37	98.8	5.93	441.4	8.94	122.6	6.12
−50	441.8	8.76	94.7	6.38	387.5	9.36	117.8	6.59
−40	389.8	9.16	90.7	6.84	342.6	9.79	113.1	7.09
−30	346.2	9.55	86.9	7.32	304.6	10.21	108.5	7.61
−26	330.6	9.71	85.3	7.51	291.0	10.38	106.6	7.83
−20	309.0	9.95	83.1	7.80	271.9	10.63	103.9	8.17
−16	295.6	10.10	81.6	8.01	260.1	10.80	102.1	8.40
−10	276.9	10.34	79.4	8.31	243.4	11.06	99.3	8.77
−6	265.2	10.50	78.0	8.52	233.0	11.24	97.5	9.02
−4	259.6	10.58	77.3	8.62	227.9	11.32	96.6	9.15
−2	254.1	10.66	76.6	8.73	233.0	11.41	95.7	9.28
0	248.7	10.74	75.9	8.84	218.2	11.50	94.8	9.42
2	243.5	10.82	75.1	8.95	213.5	11.59	93.9	9.56
4	238.4	10.90	74.4	9.06	208.9	11.68	93.1	9.70
6	233.4	10.98	73.7	9.17	204.4	11.77	92.2	9.84
10	223.8	11.15	72.3	9.39	195.7	11.96	90.4	10.14
16	210.1	11.40	70.2	9.74	183.2	12.24	87.7	10.61
20	201.4	11.57	69.9	9.98	175.3	12.43	85.9	10.95
26	189.0	11.83	66.8	10.36	163.9	12.74	83.2	11.49
30	181.1	12.01	65.4	10.62	156.7	12.95	81.4	11.89
36	169.8	12.28	63.4	11.03	146.1	13.28	78.7	12.54
40	162.5	12.48	62.1	11.33	139.4	13.52	76.9	13.02
46	152.0	12.78	60.0	11.79	129.5	13.90	74.1	13.83
50	145.3	12.99	58.7	12.13	123.1	14.18	72.3	14.45
60	129.1	13.57	55.3	13.08	107.6	14.98	67.6	16.36
70	113.6	14.26	52.0	14.25	92.4	16.02	62.9	19.16
80	98.6	15.11	48.7	15.80	76.6	17.55	58.6	23.87
85	91.1	15.65	47.2	16.82	68.1	18.71	57.4	27.82
90	83.4	16.29	45.9	18.11	58.3	20.48	59.3	34.55

$\mu = Ns/m^2$ $k = W/mK$

Table 6: Variation of Thermal Conductivity and Viscosity of R134a, Tetrafluoroethane and R717, Ammonia

Temperature °C	R134a, Tetrafluoroethane Viscosity, $\mu \times 10^6$	R134a, Tetrafluoroethane Conductivity, $k \times 10^3$		R717, Ammonia Viscosity, $\mu \times 10^6$	R717, Ammonia Conductivity, $k \times 10^3$	
−70	809.2	7.89	126.0	5.75	475.0	7.03	792.1	19.73
−60	663.1	8.30	120.7	6.56	391.3	7.30	757.0	19.93
−50	535.1	8.72	115.6	7.36	328.9	7.57	722.3	20.24
−40	482.2	9.12	110.6	8.17	281.2	7.86	688.1	20.64
−30	406.4	9.52	105.8	8.99	244.1	8.15	654.6	21.15
−26	383.8	9.68	103.9	9.32	231.4	8.27	641.5	21.38
−20	353.0	9.92	101.1	9.82	214.4	8.45	622.0	21.77
−16	344.3	10.09	99.2	10.15	204.2	8.57	609.1	22.05
−10	308.6	10.33	96.5	10.66	190.2	8.75	590.1	22.50
−6	292.9	10.49	94.7	11.00	181.7	8.87	577.7	22.83
−4	285.4	10.57	93.8	11.17	177.7	8.93	571.5	23.00
−2	278.1	10.65	92.9	11.34	173.8	8.99	565.3	23.18
0	271.1	10.73	92.0	11.51	170.1	9.06	559.2	23.37
2	264.3	10.81	91.1	10.69	166.5	9.12	553.1	23.55
4	257.6	10.90	90.2	11.86	162.9	9.18	547.1	23.75
6	251.2	10.98	89.4	12.04	159.5	9.24	541.1	23.95
10	238.8	11.15	87.6	12.40	153.0	9.36	529.1	24.37
16	221.5	11.40	85.0	12.95	144.0	9.55	511.5	25.04
20	210.7	11.58	83.3	13.33	138.3	9.68	499.9	25.52
26	195.4	11.85	80.7	13.92	130.4	9.87	482.7	26.29
30	185.8	12.04	79.0	14.33	125.5	10.00	471.4	26.85
36	172.1	12.34	76.4	14.98	118.4	10.19	454.6	27.74
40	163.4	12.55	74.7	15.44	114.0	10.33	443.5	28.38
46	151.0	12.88	72.1	16.18	107.8	10.53	427.1	29.41
50	143.1	13.12	70.4	16.72	103.8	10.67	416.3	30.16
60	124.2	13.79	66.1	18.31	94.5	11.05	389.6	32.26
70	106.4	14.65	61.7	20.45	85.9	11.47	363.2	34.80
80	89.0	15.84	57.2	23.72	78.0	11.95	337.1	38.00
85	80.2	16.67	54.9	26.22	74.2	12.23	324.1	39.95
90	70.9	17.81	52.8	29.91	70.5	12.55	311.0	42.24

$\mu = Ns/m^2$ $k = W/mK$

Table 7: Variation of Thermal Conductivity and Viscosity of R744, Carbon Dioxide and R32, Difluoromethane

Temperature °C	Refrigerant 744, Carbon Dioxide Viscosity, $\mu \times 10^6$ Liquid	Viscosity, $\mu \times 10^6$ Vapour	Th. Conductivity, $k \times 10^3$ Liquid	Th. Conductivity, $k \times 10^3$ Vapour	Refrigerant 32, Difluoromethane Viscosity, $\mu \times 10^6$ Liquid	Viscosity, $\mu \times 10^6$ Vapour	Th. Conductivity, $k \times 10^3$ Liquid	Th. Conductivity, $k \times 10^3$ Vapour
−40	193.8	11.87	159.3	12.54	244.6	9.75	182.2	8.93
−38	187.4	11.98	156.8	12.75	239.0	9.84	180.8	9.05
−32	169.7	12.34	149.3	13.43	223.2	10.10	176.7	9.40
−28	158.9	12.59	144.4	13.94	213.4	10.27	173.9	9.65
−24	148.8	12.85	139.5	14.49	204.1	10.45	171.1	9.91
−20	139.3	13.12	134.6	15.09	195.2	10.63	168.3	10.18
−18	134.8	13.26	132.2	15.42	191.0	10.73	166.9	10.32
−14	126.2	13.55	127.4	16.14	182.7	10.91	164.1	10.61
−12	122.0	13.70	125.0	16.54	178.8	11.01	162.7	10.75
−10	118.0	13.86	122.5	16.96	174.9	11.10	161.3	10.92
−8	114.1	14.03	120.1	17.42	171.1	12.20	159.8	11.08
−6	110.3	14.20	117.7	17.91	167.4	11.30	158.4	11.25
−4	106.6	14.39	115.3	18.44	163.8	11.40	157.0	11.42
−2	102.9	14.58	112.9	19.03	160.2	11.50	155.5	11.60
0	99.4	14.79	110.4	19.67	156.7	11.60	154.1	11.79
2	95.9	15.00	108.0	20.38	153.3	11.71	152.6	11.98
4	92.5	15.24	105.5	21.17	149.9	11.81	151.1	12.19
6	89.1	15.49	103.1	22.06	146.7	11.92	149.6	12.40
8	85.8	15.76	100.6	23.06	143.4	12.03	148.2	12.62
10	82.6	16.06	98.1	24.21	140.3	12.14	146.7	12.86
12	79.3	16.39	95.6	25.53	137.1	12.25	145.1	13.11
14	76.1	16.75	93.1	27.08	134.1	12.37	143.6	13.38
16	72.8	17.16	90.6	28.93	131.1	12.48	142.1	13.67
18	69.5	17.64	88.1	31.16	128.1	12.60	140.5	13.37
20	66.1	18.19	85.7	33.94	125.2	12.73	139.0	14.30
22	62.7	18.85	83.4	37.52	122.3	12.85	137.4	14.65
24	59.0	19.66	81.5	42.35	119.5	12.98	135.8	15.03
26	55.0	20.73	80.5	49.44	116.7	13.11	134.2	15.44
28	50.3	22.27	81.9	61.73	113.9	13.25	132.6	15.88
30	43.8	25.17	95.4	98.02	111.2	13.39	131.0	16.36

$\mu = Ns/m^2$ $k = W/mK$

Table 8: Transport Properties of Water and Steam at Saturation Conditions

t, °C	P, bar	Specific heat c_p, kJ/kg K Liquid	Specific heat c_p, kJ/kg K Vapour	Dynamic Viscosity, μ, Ns/m² × 10⁻⁶ Liquid	Dynamic Viscosity, μ, Ns/m² × 10⁻⁶ Vapour	Thermal Conductivity, k, W/mK × 10⁻³ Liquid	Thermal Conductivity, k, W/mK × 10⁻³ Vapour	Prandtl Number Liquid	Prandtl Number Vapour
60	.1992	4.185	1.915	466.8	10.94	650.7	21.10	3.002	.9926
80	.4736	4.196	1.962	355.0	11.60	666.8	22.86	2.234	.9951
100	1.013	4.216	2.028	282.2	12.28	677.5	24.79	1.756	1.004
105	1.208	4.222	2.048	267.9	12.45	679.3	25.31	1.665	1.007
110	1.433	4.229	2.070	254.9	12.62	680.9	25.84	1.583	1.011
115	1.691	4.236	2.094	243.0	12.79	682.2	26.39	1.509	1.015
120	1.985	4.245	2.120	232.1	12.97	683.3	26.96	1.442	1.020
125	2.321	4.254	2.147	222.0	13.14	684.0	27.55	1.380	1.024
130	2.701	4.263	2.176	212.7	13.32	684.4	28.15	1.325	1.030
135	3.131	4.274	2.207	204.1	13.49	684.6	28.77	1.274	1.035
140	3.614	4.285	2.241	196.1	13.67	684.5	29.42	1.228	1.041
145	4.155	4.297	2.276	188.7	13.84	684.2	30.09	1.185	1.047
150	4.760	4.310	2.314	181.9	14.02	683.6	30.77	1.147	1.054
155	5.433	4.324	2.355	175.5	14.19	682.7	31.49	1.112	1.062
160	6.181	4.339	2.397	169.5	14.37	681.5	32.22	1.079	1.069
165	7.008	4.354	2.443	164.0	14.55	680.1	32.98	1.050	1.077
170	7.920	4.371	2.491	158.8	14.72	678.5	33.77	1.023	1.086
180	10.03	4.408	2.596	149.3	15.07	674.5	35.42	.9760	1.104
190	12.55	4.449	2.713	141.0	15.42	669.4	37.20	.9372	1.125
200	15.55	4.497	2.843	133.6	15.78	663.4	39.10	.9056	1.147
210	19.08	4.551	2.988	127.0	16.13	656.3	41.14	.8802	1.172
220	23.20	4.613	3.150	121.0	16.49	648.3	43.35	.8607	1.198
230	27.98	4.685	3.331	115.5	16.85	639.3	45.74	.8464	1.227
240	33.48	4.769	3.536	110.5	17.22	629.2	48.34	.8373	1.260
250	39.78	4.867	3.772	105.8	17.59	618.1	51.18	.8333	1.297
260	46.94	4.983	4.047	101.5	17.98	605.9	54.33	.8346	1.339
270	55.06	5.122	4.373	97.36	18.38	592.6	57.84	.8415	1.390
280	64.20	5.290	4.767	93.41	18.80	578.0	61.82	.8550	1.450
290	74.46	5.499	5.253	89.58	19.25	562.2	66.40	.8762	1.523
300	85.93	5.762	5.863	85.82	19.74	545.0	71.78	.9072	1.612
310	98.70	6.104	6.650	82.07	20.28	526.4	78.26	.9516	1.723
320	112.9	6.565	7.722	78.27	20.89	506.3	86.34	1.015	1.868
330	128.6	7.219	9.361	74.36	21.62	484.5	96.93	1.108	2.088
340	146.1	8.233	12.21	70.23	22.52	461.1	111.8	1.254	2.460
350	165.4	10.10	17.15	65.66	23.75	436.3	134.6	1.521	3.024
360	186.8	14.58	25.12	60.11	25.56	411.8	176.8	2.128	3.631
370	210.5	43.17	76.91	51.82	29.31	416.4	306.4	5.372	7.358

Table 9: Thermal Conductivity, k, of Steam and Water at Various Pressures and Temperatures

P, bar t °C	1	5	10	25	50	75	100	150	200
60	.6507	.6509	.6512	.6520	.6533	.6546	.6559	.6585	.6610
80	**.6668**	.6670	.6673	.6681	.6694	.6708	.6721	.6748	.6774
100	.02478	**.6777**	.6780	.6788	.6802	.6817	.6831	.6858	.6885
120	.02629	**.6834**	.6838	.6847	.6862	.6877	.6892	.6921	.6950
140	.02794	**.6846**	.6850	.6860	.6876	.6892	.6908	.6940	.6971
160	.02968	.03142	**.6818**	.6829	.6847	.6865	.6832	.6916	.6950
180	.03149	.03263	.03541	**.6757**	.6776	.6796	.6815	.6853	.6890
200	.03337	.03424	.03606	**.6642**	.6665	.6687	.6708	.6750	.6792
220	.03529	.03598	.03728	**.6485**	.6511	.6536	.6561	.6609	.6655
240	.03727	.03784	.03883	.04381	**.6312**	.6342	.6370	.6426	.6480
260	.03930	.03979	.04059	.04424	**.6063**	.6099	.6134	.6200	.6263
280	.04137	.04181	.04249	.04534	.05370	**.5799**	.5843	.5925	.6001
300	.04349	.04389	.04449	.04683	.05304	.06406	**.5482**	.5589	.5686
320	.04565	.04603	.04656	.04856	.05348	.06132	.07472	**.5173**	.5303
340	.04785	.04821	.04870	.05047	.05454	.06057	.06980	**.4629**	.4827
360	.05010	.05044	.05089	.05248	.05595	.06080	.06775	.09567	**.4195**
380	.05238	.05271	.05314	.05459	.05759	.06150	.06668	.08397	.1296
400	.05471	.05502	.05543	.05679	.05953	.06295	.06726	.07996	.1034
420	.05707	.05737	.05777	.05905	.06157	.06461	.06832	.07847	.09465
440	.05947	.05976	.06014	.06137	.06371	.06647	.06975	.07831	.09081
460	.06191	.06219	.06256	.06373	.06593	.06848	.07145	.07891	.08918
480	.06438	.06466	.06501	.06613	.06822	.07061	.07333	.07999	.08878
500	.06689	.06716	.06750	.06858	.07057	.07282	.07535	.08140	.08913
520	.06943	.06969	.07002	.07107	.07298	.07511	.07748	.08306	.09000
540	.07200	.07225	.07258	.07359	.07543	.07746	.07970	.08489	.09122
560	.07460	.07485	.07516	.07615	.07792	.07986	.08199	.08687	.09271
580	.07723	.07747	.07778	.07874	.08045	..08232	.08435	.08896	.09439
600	.07989	.08013	.08043	.08136	.08301	.08481	.08676	.09114	.09624
620	.08258	.08281	.08310	.08400	.08561	.08735	.08922	.09347	.09822
640	.08529	.08551	.08580	.08668	.08824	.08993	.09173	.09573	.1003
660	.08812	.08824	.08852	.08938	.09090	.09253	.09428	.09812	.1025
680	.09078	.09099	.09127	.09211	.09359	.09517	.09686	.1006	.1047
700	.09356	.09377	.09403	.09486	.09630	.09784	.09947	.1030	.1070

Note: Steam upto bold character and water after that.

Table 10: Dynamic Viscosity, ($\mu \times 10^{-6}$) of Steam and Water at Various Pressures and Temperatures

P, bar t °C	1	5	10	25	50	75	100	150	200
60	466.8	466.9	467.0	467.3	467.9	468.4	468.9	470.0	471.1
80	**355.0**	355.1	355.2	355.6	356.2	356.9	357.5	358.8	360.1
100	12.28	**282.3**	282.4	282.8	283.5	284.2	284.9	286.2	287.5
120	13.03	**232.1**	232.3	232.7	233.4	234.0	234.7	236.0	237.4
140	13.80	**196.2**	196.3	196.7	197.4	198.0	198.7	200.0	201.3
160	14.58	14.42	**169.6**	170.0	170.7	171.3	172.0	173.2	174.5
180	15.38	15.24	15.07	**149.7**	150.4	151.0	151.6	152.9	154.1
200	16.18	16.07	15.93	**133.6**	134.5	135.1	135.7	137.0	138.2
220	16.99	16.90	16.79	**121.0**	121.7	122.3	122.9	124.2	125.4
240	17.81	17.74	17.64	17.37	**110.9**	111.6	112.3	113.6	114.8
260	18.63	18.57	18.50	18.28	**101.6**	102.3	103.0	104.4	105.8
280	19.46	19.41	19.35	19.17	18.92	**93.78**	94.61	96.19	97.68
300	20.29	20.25	20.20	20.06	19.86	19.74	**86.39**	88.30	90.05
320	21.12	21.09	21.05	20.94	20.79	20.71	20.76	**80.17**	82.40
340	21.95	21.93	21.90	21.81	21.70	21.66	21.71	**70.56**	74.05
360	22.79	22.77	22.74	22.68	22.60	22.59	22.65	23.25	**62.57**
380	23.62	23.60	23.58	23.54	23.50	23.50	23.58	24.05	25.68
400	24.45	24.44	24.43	24.40	24.38	24.40	24.49	24.91	25.96
420	25.28	25.27	25.26	25.25	25.25	25.30	25.39	25.79	26.61
440	26.11	26.10	26.10	26.09	26.12	26.18	26.28	26.66	27.37
460	26.93	26.93	26.93	26.94	26.98	27.05	27.16	27.54	28.18
480	27.75	27.76	27.76	27.78	27.83	27.91	28.03	28.41	29.00
500	28.57	28.58	28.58	28.61	28.67	28.77	28.90	29.27	29.82
520	29.39	29.40	29.40	29.44	29.51	29.62	29.75	30.12	30.65
540	30.20	30.21	30.22	30.26	30.34	30.46	30.60	30.97	31.48
560	31.01	31.02	31.03	31.08	31.17	31.29	31.43	31.80	32.30
580	31.81	31.83	31.84	31.89	31.99	32.12	32.26	32.63	33.11
600	32.62	32.63	32.64	32.70	32.81	32.94	33.09	33.46	33.93
620	33.41	33.43	33.44	33.50	33.62	33.75	33.90	34.27	34.73
640	34.20	34.22	34.24	34.30	34.42	34.55	34.71	35.08	35.53
660	34.99	35.01	35.03	35.09	35.21	35.35	35.51	35.88	36.32
680	35.77	35.79	35.81	35.88	36.01	36.15	36.31	36.67	37.11
700	36.55	36.57	36.59	36.66	36.79	36.94	37.09	37.46	37.89

Note: Steam upto bold character and water after that.

Table 11: Prandtl Number of Steam and Water at Various Pressures and Temperatures

P, bar t °C	1	5	10	25	50	75	100	150	200
60	3.002	3.001	3.000	2.996	2.989	2.983	2.976	2.964	2.953
80	**2.234**	2.233	2.233	2.231	2.227	2.224	2.221	2.215	2.209
100	1.004	**1.756**	1.755	1.754	1.752	1.751	1.749	1.746	1.743
120	0.994	**1.441**	1.441	1.441	1.439	1.438	1.437	1.436	1.434
140	0.983	**1.228**	1.227	1.227	1.226	1.225	1.225	1.223	1.222
160	0.974	1.051	**1.079**	1.079	1.078	1.077	1.076	1.075	1.073
180	0.966	1.033	1.104	**0.975**	0.974	0.973	0.972	0.970	0.969
200	0.960	1.014	1.081	**0.905**	0.903	0.902	0.900	0.898	0.895
220	0.954	0.997	1.054	**0.860**	0.858	0.856	0.854	0.850	0.846
240	0.949	0.983	1.028	1.176	**0.835**	0.831	0.828	0.822	0.817
260	0.945	0.971	1.006	1.130	**0.834**	0.828	0.823	0.813	0.805
280	0.941	0.961	0.988	1.086	1.289	**0.851**	0.841	0.825	0.812
300	0.938	0.953	0.974	1.050	1.211	1.444	**0.897**	0.866	0.843
320	0.935	0.947	0.962	1.020	1.147	1.310	1.582	**0.962**	0.912
340	0.932	0.941	0.953	0.997	1.095	1.215	1.382	**1.233**	1.068
360	0.929	0.936	0.946	0.981	1.055	1.147	1.259	1.662	**1.675**
380	0.926	0.932	0.939	0.966	1.025	1.099	1.186	1.451	2.021
400	0.924	0.928	0.934	0.955	1.000	1.057	1.123	1.298	1.626
420	0.921	0.925	0.930	0.946	0.981	1.025	1.078	1.204	1.394
440	0.919	0.922	0.926	0.939	0.966	1.001	1.042	1.139	1.264
460	0.916	0.919	0.922	0.933	0.955	0.982	1.015	1.092	1.182
480	0.913	0.916	0.919	0.928	0.946	0.967	0.993	1.055	1.126
500	0.911	0.913	0.916	0.924	0.938	0.956	0.977	1.026	1.083
520	0.908	0.910	0.913	0.920	0.933	0.947	0.964	1.004	1.050
540	0.906	0.908	0.910	0.916	0.927	0.940	0.953	0.986	1.023
560	0.903	0.905	0.907	0.913	0.923	0.934	0.945	0.972	1.002
580	0.901	0.902	0.904	0.910	0.919	0.928	0.938	0.960	0.985
600	0.898	0.900	0.902	0.907	0.915	0.924	0.933	0.951	0.972
620	0.896	0.897	0.899	0.904	0.912	0.920	0.928	0.944	0.961
640	0.893	0.895	0.896	0.901	0.909	0.916	0.923	0.937	0.952
660	0.891	0.892	0.894	0.898	0.905	0.912	0.919	0.932	0.945
680	0.889	0.890	0.891	0.895	0.902	0.909	0.915	0.927	0.938
700	0.886	0.887	0.889	0.893	0.899	0.905	0.911	0.922	0.933

Note: Steam upto bold font and then water.

UNIT CONVERSION CONSTANTS

Quantity	SI to Metric	Metric to SI
Force	1 N = 0.1019 kg_f	1 kg_f = 9.81 N
Pressure	1 N/m^2 = 10.19 × 10^{-6} kg_f/cm^2	1 kg_f/cm^2 = 98135 N/m^2, (Pascal)
	1 bar = 1.0194 kg_f/cm^2	1 kg_f/cm^2 = 0.9814 bar
Energy	1 kJ = 0.2389 kcal	1 kcal = 4.186 kJ
(heat, work)	1 Nm (= 1J) = 0.1019 kg_f m	1 kg_f m = 9.81 Nm, (J)
	1 kW hr = 1.36 hp hr	1 hp hr = 0.736 kW hr
Power (metric)	1 W = 1.36 × 10^{-3} hp	1 hp = 736 W
Heat flow	1 W = 0.86 kcal/hr	1 kcal/hr = 1.163 W
Specific heat	1 kJ/kg K = 0.2389 kcal/kg °C	1 kcal/kg °C = 4.186 kJ/kg K
Surface Tension	1 N/m = 0.1019 kg_f/m	1 kg_f/m = 9.81 N/m
Thermal Conductivity	1 W/m K = 0.86 kcal/hr m °C	1 kcal/hr m °C = 1.163 W/m K
Convection Coefficient	1 W/m^2 K = 0.86 kcal/hr m^2 °C	1 kcal/hr m^2 °C = 1.163 W/m^2 K
Dynamic Viscosity	1 kg/ms, (Ns/m^2) = 0.1 Poise	1 Poise = 10 kg/ms, (Ns/m^2), Pa s
Kinematic Viscosity	1 m^2/s = 3600 m^2/hr,	1 m^2/hr = 2.778 × 10^{-4} m^2/s
	1 Stoke = 10^{-4} m^2/s	

Universal gas constant = 8314.41 J/kg mol K = 847.54 mkg_f/kg mol K = 1.986 kcal/kg mol K

Gas Constant for Air = 287 J/kg K, c_p = 1005 J/kg K = 0.24 kcal/kg K

Stefan-Boltzmann constant = 5.67 × 10^{-8} W/m^{2}K^4 = 4.876 × 10^{-8} kcal/hr m^2 K^4

REFERENCES

1. Eckert, E.R.G. and Drake, R.M. (1959). *Heat and Mass Transfer*, McGraw-Hill.
2. Holman, P.J. (2002, 9th edition). *Heat Transfer*, McGraw-Hill.
3. Incropera, J.P. and Dewitt, D.P. (2002). Fourth edition, *Introduction to Heat Transfer*. John Wiley and Sons.
4. Kern, D.Q. (1950). *Process Heat Transfer,* McGraw-Hill.
5. Kreith, F. and Bohn, M.S. (1997). *Principles of Heat Transfer*, Fifth edition, PWS Publishing Co., an International Thomson Publishing Co.
6. Kutateladze, S.S. and Borishankii, V.M. (1966). *A Concise Encyclopedia of Heat Transfer*, Pergamon Press.
7. McAdams, W.H. (1954). *Heat Transmission*, McGraw-Hill.
8. Rohsenow, W.M. et al. (1985). *Hand Book of Heat Transfer Applications and Hand Book of Heat Transfer Fundamentals*, McGraw-Hill.
9. Schneider, P.J. (1957). *Conduction Heat Transfer*, Addison-Wesley Publishing Co.
10. Spalding D.B. (1963). *Convective Mass Transfer : An Introduction*, Edward Arnold.
11. Streeter, V.L. (1983). *Fluid Mechanics*, McGraw-Hill.
12. Thomas, L.C. (1980). *Fundamentals of Heat Transfer*, Prentice Hill.
13. Treybal, R.E. (1995). *Mass Transfer Operations*, McGraw-Hill.
14. Wrangham, D.A. (1961). *The Elements of Heat Flow*, Chatto and Windus.
15. Mills, A.F., (1995). *Heat and Mass Transfer,* Richard. D. Irwin Inc.
16. Dunn, P.D. and Reay, D.A. (1994). *Heat Pipes*, Pergaman Press.